Jochen Braumüller

Quantum simulation experiments with superconducting circuits

Experimental Condensed Matter Physics
Band 20

Herausgeber
Physikalisches Institut
Prof. Dr. David Hunger
Prof. Dr. Alexey Ustinov
Prof. Dr. Georg Weiß
Prof. Dr. Wolfgang Wernsdorfer
Prof. Dr. Wulf Wulfhekel

Eine Übersicht aller bisher in dieser Schriftenreihe erschienenen Bände finden Sie am Ende des Buchs.

Quantum simulation experiments with superconducting circuits

by
Jochen Braumüller

KIT Scientific Publishing

Dissertation, Karlsruher Institut für Technologie
KIT-Fakultät für Physik

Tag der mündlichen Prüfung: 8. Dezember 2017
Referenten: Prof. Dr. Alexey V. Ustinov, Prof. Dr. Gerd Schön

Impressum

KIT Scientific Publishing

Karlsruher Institut für Technologie (KIT)
KIT Scientific Publishing
Straße am Forum 2
D-76131 Karlsruhe

KIT Scientific Publishing is a registered trademark
of Karlsruhe Institute of Technology.
Reprint using the book cover is not allowed.

www.ksp.kit.edu

Print on Demand 2018 – Gedruckt auf FSC-zertifiziertem Papier

ISSN 2191-9925
ISBN 978-3-7315-0780-2
DOI 10.5445/KSP/1000081315

Contents

1 Introduction

Understanding basic principles and phenomena in nature has been a perpetual motivation for researchers since the beginning of mankind. Classical computers, acting as universal computational devices in the sense of deterministic Turing machines [Tur37], provide a powerful tool to efficiently simulate the dynamics of physical systems obeying Newtonian physics. With the advent of quantum mechanics, it became however apparent that the properties and dynamics of microscopic systems in particular are not captured by classical physics in general. A quantum mechanical treatment is typically required on atomic scales, for instance in order to treat single molecules or proteins of biological systems or general quantum models in condensed matter physics. These quantum problems have been proven incompatible with the original Turing hypothesis [Deu85] and resilient to be efficiently simulated with classical computers. However, it was proposed [Fey82; Deu85] and verified [Llo96] that any physical system can be simulated by a universal quantum computer.

In quantum information processing, the classical bit with possible states 0, 1 is replaced by the quantum bit or qubit that can assume any superposition state $|\psi\rangle = \alpha|\mathrm{g}\rangle + \beta|\mathrm{e}\rangle$, with qubit eigenstates $|\mathrm{g}\rangle$ and $|\mathrm{e}\rangle$. Due to the fundamental principle of quantum entanglement, the quantum state of N interacting qubits must be described by a common state in their joint Hilbert space of 2^N dimensions and in general cannot be decomposed into a product state of N single qubit states. In solving a quantum problem with classical hardware, the computer needs to keep track of all probability amplitudes for any possible configuration of the system at any time [GAN14], leading to an exponential explosion of the required computational power and memory.

A quantum state collapses to one of its eigenstates during a measurement process with probabilities dependent on the initial quantum state. The output of a quantum computer is therefore always classical. Many identical successive computations however yield a probability distribution that allows for the estimation of expectation values. During the computation process, several qubits interacting with each other can make use of their intrinsic quantumness. Since a quantum system can be in exponentially many states at the same time, as imposed by the superposition principle, a quantum computation is performed for all these possible configurations at once. This massive parallelism provides quantum computers with their enormous power and enables them to outperform classical computers by an exponential speed-up for some problems [Fey82; Llo96; Lan+10]. In this context, the term quantum supremacy was introduced, denoting the superiority of a quantum computation over a classical approach. Even the simulation of a quantum computer containing about 50 physical qubits, as it was announced for the near future by Google, is already at the limit of the capabilities of today's classical supercomputers [Boi+18; CZ03].

A prominent example for an exponential speed-up of quantum computers is prime factorization based on the Shor algorithm [Sho94]. This is known to be a hard problem for classical computers and it is fundamental to the RSA cryptosystem. Several proof-of-principle implementations of a compiled version of Shor's algorithm with a pre-defined small number have been demonstrated in nuclear magnetic resonance [Van+01], with cold atoms [Gil+08], on a photonic chip [PMO09], and with superconducting circuits [Luc+12].

However, the implementation of a universal quantum computer capable of performing useful calculations is very challenging since it requires many error-corrected logical qubits that in turn involve a large overhead of physical qubits. In order to obtain a single logical qubit of reasonable error rate, on the order of 10^3 to 10^4 physical qubits of present coherence rates are necessary. For factorizing a 15 bit number using Shor's algorithm, a quantum computer would require up to $\sim 10^7$ physical qubits, dependent on the tolerated error rate and the time of the computation [Fow+12].

One of the most anticipated applications of quantum computation is the simulation of quantum chemistry [Bab+15; Lan+10]. As an example, protein complexes such as ferredoxin Fe_2S_2 or the Fenna-Matthews-Olson (FMO) complex are known to mediate energy transfer in many metabolic reactions but are intractable on a classical computer [Svo14; Mos+12]. Based on the FMO complex, the efficiency of light harvesting in photosynthesis has been found to notably exceed the expectation based on classical models, such that a quantum description is likely to be required in order to understand the mechanism [Eng+07]. Astonishingly, quantum effects seem to be relevant even though these biological reactions occur at ambient temperatures. Within the electronic structure problem [Bab+15], relevant spin orbitals of active molecules are mapped to qubits, such that more than 100 logical qubits need to be taken into account in a quantum simulation solving for the ground state of Fe_2S_2 [Svo14]. Further prominent applications of quantum simulation in chemistry are the elucidation of nitrogen fixation by bacterial catalysis, the carbon capture process and material research [Rei+17].

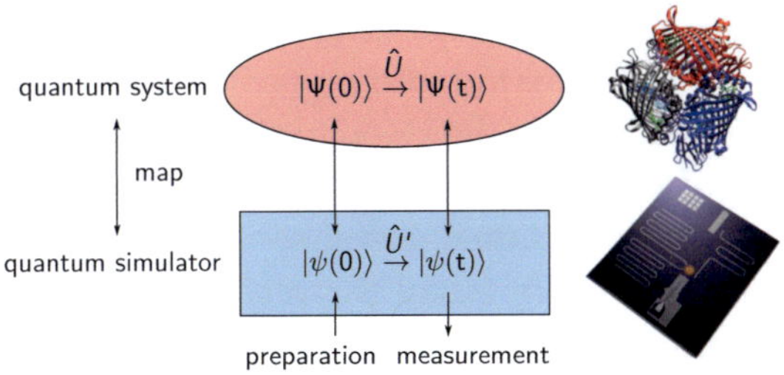

Figure 1.1: Analog quantum simulation A quantum system of interest (red), artistically depicted as a protein complex, is mapped onto an artificial quantum simulator (blue). In this thesis the quantum simulator is a superconducting circuit. By preparing and reading out quantum states $|\psi\rangle$ of the simulator system, the time evolution $\hat{U}$ of the underlying quantum problem, described by the state $|\Psi\rangle$, can be inferred [GAN14].

While the universal quantum computer is yet to be demonstrated, the concept of quantum simulation in the spirit of Feynman's original proposal [Fey82] may provide a more feasible approach to achieve quantum supremacy in the near future.

> "*Let the computer itself be built of quantum mechanical elements which obey quantum mechanical laws.*" Feynman 1982 [Fey82]

It is the key idea to implement quantum computation by using hardware that itself obeys the laws of quantum mechanics in order to avoid an exponential scaling in computational resources.

The main simplification of a quantum simulator as compared to a universal quantum computer is that the simulator is very problem-specific [GAN14]. On the other hand, this allows for the investigation of interesting quantum problems already today by combining few elements such as single qubits and resonators that are available on various experimental platforms [BN09].

The quest in quantum simulation, which comes in two flavours, is to solve the time evolution of a certain quantum system of interest.

The framework of analog quantum simulation, used in this thesis, is schematically depicted in Fig. 1.1. A quantum system of interest is mapped onto a tailored and well controllable artificial quantum system, striving to mimic the dynamics of the investigated quantum system. In the setting of this thesis, the simulator is a superconducting quantum circuit. When approximately the same equations of motion hold for both systems, the solution of the underlying quantum problem can be inferred from observing the time evolution of the artificially built model system.

The alternative approach, named digital quantum simulation, is a gate-based scheme that is in much closer resemblance to the universal quantum computer. Here, the complex unitary transformation $\hat{U}$ in Fig. 1.1 is decomposed into many single- and two-qubit gates [GAN14]. The error induced by such a Trotter decomposition arises from non-vanishing commutators of the Hamiltonians in different Trotter steps and scales with the size of individual Trotter steps [Bab+15]. As a consequence, many gates are typically required to obtain a useful result, which again can require an error correction scheme and thereby entail an overhead in resources.

First experiments on quantum simulation have been performed on various experimental platforms. Examples of analog quantum simulation are the study of fermionic transport [Sch+12], magnetism [Gre+13] and a quantum phase transition in the Bose-Hubbard model [Gre+02] with cold atoms. A simulation of the Fermi-Hubbard model was performed with an array of semiconductor quantum dots [Byr+08; Hen+17] and the simulation of a quantum magnet and the Dirac equation was demonstrated with trapped ions [Fri+08; Ger+10]. Digital simulation schemes with superconducting devices were demonstrated for fermionic models [Bar+15] and spin systems [Sal+15]. Further experiments following a digital approach are the simulation of frustrated magnets by a nuclear magnetic resonance quantum processor [Zha+12] and the dynamics of spin systems with trapped ions [Lan+11].

In this thesis, we make use of superconducting circuits that have proven to be a suitable experimental platform to implement quantum hardware [Par14; HTK12; YN11]. The non-linearity of the Josephson junction, forming the key element of all superconducting qubits, allows one to isolate two excitation levels in a quantum circuit to form a computational qubit basis [CW08; SG08]. Superconducting circuits operate at frequencies in the few gigahertz regime, much below the typical energy scale of the superconducting gap. Therewith, individual transitions are well protected from quasi-particle excitations. By cooling the quantum circuits to milli-Kelvin temperatures, thermal excitations are additionally suppressed. Superconducting qubits feature individual control, readout and frequency tunability and their properties are rather straightforward to tailor by circuit design. During almost two decades, superconducting qubits experienced a rapid improvement of their coherence properties [DS13]. This has allowed for the demonstration of several major milestones in the pursuit of scalable quantum computation, such as the control and entanglement of multiple qubits [Ste+06; Bar+14], the implementation of a quantum error correction scheme [Ree+12], the demonstration of quantum algorithms [Luc+12] and encoding quantum information in complex cavity states [Vla+13].

Here, we realize an analog quantum simulation scheme using few elements from the quantum toolbox available in superconducting circuits. This allows us to investigate quantum models that provide intriguing non-classical physics and are closely related to fundamental processes in nature. We focus in particular on the quantum Rabi model and the spin boson model that describe the interaction between matter and light on a quantum level. More specifically, the quantum Rabi model describes the interaction of a two-level atom or qubit with a quantized oscillator mode [Rab36; Rab37], while the spin boson model involves a bath of bosonic oscillator modes forming a dissipative qubit environment [Leg+87]. The more general spin boson model is ubiquitous in nature and allows one to study critical phenomena and quantum phase transitions, while finding its solution by classical means is very challenging [ABV07]. In particular for a sub-ohmic shape of the spectral function, which describes the effect of the bosonic bath on the qubit, predictions of the model dynamics at finite temperature are contradictory and a numerically exact solution does not exist [ABV07]. In addition, the spin boson model in the context of quantum simulations provides a route to experimentally study open quantum systems, for instance by engineering the dissipative vibrational environment of the electronic degrees of freedom in a biomolecule [Mos+12; Mos+16]. Recently, the spin boson model has attracted large experimental interest [Hae+15; For+16a; Mag+18; Pot+18].

The spin boson model and related quantum models exhibit their most fascinating physics in a regime where the energy associated with the coupling strength between qubit and oscillator modes becomes comparable to the subsystem energies, the so-called ultra-strong coupling regime [Cas+10]. While it is experimentally challenging to achieve ultra-strong coupling merely by sample design [Nie+10; Yos+16], and some types of qubits even seem to refuse to be pushed into such a regime [Jaa+16], we demonstrate an analog quantum simulation scheme that creates an effective quantum Rabi model in the ultra-strong coupling regime [Bal+12; Bra+17]. This simplified version of the spin boson model, involving only one single quantized oscillator mode that couples to the qubit, already reveals remarkable quantum features such as distinct collapse-revival dynamics. With this proof-of-principle experiment, we validate the experimental feasibility of simulating non-trivial quantum models with superconducting qubits.

In this thesis, I first provide the theoretical concepts necessary to understand how to operate superconducting qubits in the coherent quantum regime. The following chapter briefly elucidates the basic methods and techniques that we use to carry out the experiments in the microwave regime. In the subsequent chapter I introduce the concentric transmon qubit [Bra+16], which was conceived in collaboration with co-workers from the National Institute of Standards and Technology (NIST) in Boulder, Colorado as a robust and versatile building block for quantum simulation experiments. Thereafter, the quantum simulation of the Rabi model in the ultra-strong coupling regime is presented [Bra+17]. The last chapter details our efforts and first steps in studying the spin boson model [Lep+17]. We validate the experimental feasibility of a modular approach where qubit and bosonic bath are fabricated on two separate chips. By making use of the specific symmetry properties of the concentric transmon, we demonstrate the interaction between a single qubit and a bosonic environment that is formed by 20 individual resonators with a tailored non-trivial spectral function.

2 Quantum devices based on superconducting circuits

In this chapter, I provide the main building blocks of superconducting quantum circuits. The quantum bit as the basic unit cell of quantum devices is introduced, together with a brief discussion of its fundamental behaviour and its dissipative dynamics. Subsequently, I provide the basic concepts of superconductivity and the Josephson effect, which are the key ingredients to all superconducting quantum circuits. Hereafter, I present the working principle and the main properties of the transmon qubit, which is the type of qubit used in this thesis. The chapter concludes with an overview on cavity quantum electrodynamics and its circuit implementation.

2.1 The quantum bit

The quantum bit or qubit is a quantum mechanical two-level system with logical states $|\mathrm{g}\rangle$, $|\mathrm{e}\rangle$, in analogy to a classical bit. While a classical bit can be in only one of its fundamental states, a qubit can be in any superposition state

$$|\psi\rangle = \alpha\,|\mathrm{g}\rangle + \beta\,|\mathrm{e}\rangle \tag{2.1}$$

prior to a quantum measurement. α and β are normalized complex coefficients and $|\mathrm{g}\rangle$, $|\mathrm{e}\rangle$ denote the qubit groundstate and excited state, respectively. Since $|\mathrm{g}\rangle$, $|\mathrm{e}\rangle$ span a two-dimensional basis, they can be represented in terms of eigenvectors of the Pauli matrix $\hat{\sigma}_z$,

$$|\mathrm{g}\rangle = \begin{pmatrix} 0 \\ 1 \end{pmatrix} \quad |\mathrm{e}\rangle = \begin{pmatrix} 1 \\ 0 \end{pmatrix}, \tag{2.2}$$

with $\hat{\sigma}_z\,|g\rangle = -\,|g\rangle$ and $\hat{\sigma}_z\,|e\rangle = |e\rangle$.

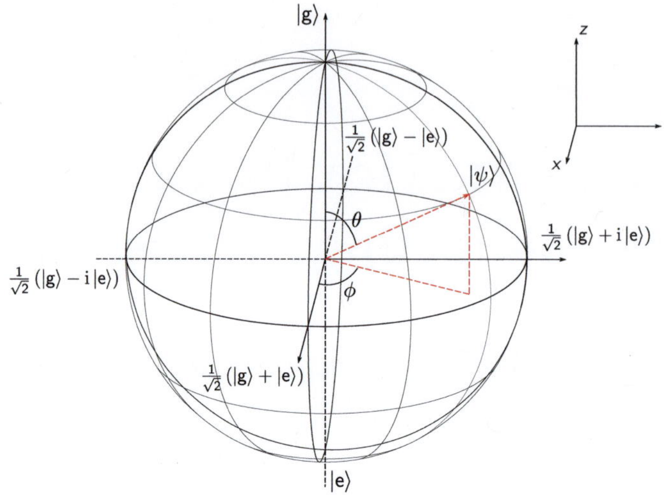

Figure 2.1: Representation of a qubit state on the Bloch sphere An arbitrary pure qubit state $|\psi\rangle$ is represented as a point on the surface of the Bloch sphere, defining the Bloch vector. It is unambiguously defined by the Euler angles ϕ and θ. The logical basis states $|\mathrm{g}\rangle$ and $|\mathrm{e}\rangle$ of the qubit are located on the poles of the Bloch sphere and are eigenvectors of $\hat{\sigma}_z$. Since $|\mathrm{g}\rangle$ and $|\mathrm{e}\rangle$ form a basis, $|\langle\psi|\psi\rangle|^2 = 1$ and the Bloch sphere is of radius 1.

2.1.1 Bloch sphere representation

It is intuitive to represent a single pure qubit state $|\psi\rangle$ as a point on the surface of the Bloch sphere, see Fig. 2.1. Here, a general qubit state is characterized by a Bloch vector with the Euler angles ϕ and θ [WS06] and Eq. (2.1) takes the form

$$|\psi\rangle = \cos\frac{\theta}{2}|\mathrm{g}\rangle + \sin\frac{\theta}{2}e^{\mathrm{i}\phi}|\mathrm{e}\rangle\,. \tag{2.3}$$

The time evolution of a general pure qubit state can be found according to

$$|\psi(t)\rangle = \hat{U}(t-t_0)\,|\psi(t_0)\rangle\,, \tag{2.4}$$

with $\hat{U} = \exp -\mathrm{i}\hat{H}(t-t_0)/\hbar$ the time evolution operator and

$$\hat{H} = \hbar\frac{\epsilon}{2}\hat{\sigma}_z \tag{2.5}$$

the time-independent qubit Hamiltonian, quantized along the z-axis. ϵ denotes the qubit energy splitting, or transition frequency. Evaluating Eq. (2.4) for $t_0 = 0$ yields

$$|\psi(t)\rangle = \cos\frac{\theta}{2}|\mathrm{g}\rangle + \sin\frac{\theta}{2}e^{\mathrm{i}(\phi-\epsilon t)}|\mathrm{e}\rangle \tag{2.6}$$

where the global phase was chosen arbitrarily. Equation 2.6 implies a phase evolution with constant angular frequency ϵ, corresponding to a steady Larmor precession around the quantization axis (z-axis). For an intuitive visualization of qubit states on the Bloch sphere, this Larmor precession is often disregarded by looking at the qubit Bloch vector from a frame rotating with the Larmor frequency ϵ.

The Bloch sphere representation can also capture mixed states, denoting a statistical mixture of several quantum states. In practice, this is used to describe an ensemble of spins interacting with a noisy environment. In qubit experiments, a mixed state can contain the statistical state of a single qubit subject to decoherence that is successively measured multiple times after a certain equal preparation and evolution. The mixed state of an ensemble is described by the density matrix [Sak94]

$$\hat{\rho} = \sum_i p_i \left|\psi_i\right\rangle \left\langle\psi_i\right| \quad \sum_i p_i = 1, \tag{2.7}$$

where p_i is the non-negative probability to find the ensemble in the pure state $|\psi_i\rangle$. Mixed states are represented by points located inside the Bloch sphere.

2.1.2 Qubit dynamics

The dynamical behaviour of a mixed qubit state is most evidently described by the master equation in Lindblad form [Lin76; HR06; BCR11],

$$\frac{\mathrm{d}}{\mathrm{d}t}\hat{\rho} = -\frac{\mathrm{i}}{\hbar}\left[\hat{H},\hat{\rho}\right] + \sum_{j=1}^{3} \Gamma_j \left(\hat{L}_j\hat{\rho}\hat{L}_j^{\dagger} - \frac{1}{2}\left[\hat{L}_j^{\dagger}\hat{L}_j,\hat{\rho}\right]_+\right) \tag{2.8}$$

Equation (2.8) yields the time evolution of the density matrix $\hat{\rho}$ in the presence of decoherence, quantified by the Lindblad or collapse operators $\hat{L}_j$ and the respective rate Γ_j for decoherence channel j. We also made use of the anticommutator notation $[a,b]_+ = ab + ba$. The set of Lindblad operators $\hat{L}_j$ are required to be traceless and orthonormal [BCR11].

The relevant decoherence mechanisms are energy relaxation at a rate Γ_1 via the qubit decay operator $\hat{L}_1 = \hat{\sigma}_-$ and pure dephasing at a rate Γ_τ via $\hat{L}_2 = \hat{\sigma}_z/\sqrt{2}$, inducing rotations around the z-axis. We define $\hat{\sigma}_- = 1/2\left(\hat{\sigma}_x - \mathrm{i}\hat{\sigma}_y\right)$ with $\hat{\sigma}_x$, $\hat{\sigma}_y$ Pauli operators and $\hat{\sigma}_- \left|\mathrm{e}\right\rangle = \left|\mathrm{g}\right\rangle$. The third decoherence channel describes qubit excitation via $\hat{\sigma}_+$ and can be neglected at very cold temperatures. Since $\mathrm{tr}\,\hat{\rho} = 1$, we can rewrite the density matrix in its generic form

$$\hat{\rho} = \begin{pmatrix} \rho_{11} & \rho_{10} \\ \rho_{10}^{*} & 1-\rho_{11} \end{pmatrix} \tag{2.9}$$

with only two independent components [Gra13]. The Lindblad equation, Eq. (2.8), becomes

$$\begin{aligned}\frac{\mathrm{d}}{\mathrm{d}t}\begin{pmatrix}\rho_{11} & \rho_{10}\\ \rho_{10}^{*} & 1-\rho_{11}\end{pmatrix} &= -\frac{\mathrm{i}}{\hbar}\left[\hat{H},\hat{\rho}\right] + \begin{pmatrix}-\Gamma_1\rho_{11} & -(\Gamma_1/2+\Gamma_\tau)\rho_{10}\\ -(\Gamma_1/2+\Gamma_\tau)\rho_{10}^{*} & \Gamma_1\rho_{11}\end{pmatrix}\\ &= -\frac{\mathrm{i}}{\hbar}\left[\hat{H},\hat{\rho}\right] + \begin{pmatrix}-\Gamma_1\rho_{11} & -\Gamma_2^{*}\rho_{10}\\ -\Gamma_2^{*}\rho_{10}^{*} & \Gamma_1\rho_{11}\end{pmatrix}.\end{aligned} \tag{2.10}$$

In the last step we introduced the dephasing rate

$$\Gamma_2^{*} = \frac{1}{2}\Gamma_1 + \Gamma_\tau \tag{2.11}$$

according to the common convention with an asterisk. The symbol Γ_2 without the asterisk usually denotes the dephasing rate acquired with a Hahn echo pulse in the measurement sequence that eliminates low-frequency noise [Hah50], see Sec. 5.5.

In the absence of pure dephasing where $1/\Gamma_\tau \equiv \tau = \infty$, it is apparent from Eq. (2.11) that the dephasing time $T_2^{*} = 1/\Gamma_2^{*}$ is relaxation limited,

$$T_2^{*} = 2T_1. \tag{2.12}$$

Using the qubit Hamiltonian defined in Eq. (2.5) but transformed into the frame rotating at a frequency ω_1,

$$\hat{H}_{\mathrm{rf}} = \hbar\frac{\epsilon - \omega_1}{2}\hat{\sigma}_z = \hbar\frac{\Delta\omega}{2}\hat{\sigma}_z, \tag{2.13}$$

allows us to solve Eq. (2.10) for $\hat{\rho}(t)$, which yields

$$\hat{\rho}(t) = \begin{pmatrix} c_0 e^{-\Gamma_1 t} & c_1 e^{-\Gamma_2^{*} t} e^{-\mathrm{i}\Delta\omega t}\\ c_1^{*} e^{-\Gamma_2^{*} t} e^{\mathrm{i}\Delta\omega t} & 1 - c_0 e^{-\Gamma_1 t}\end{pmatrix} \tag{2.14}$$

with integration constants c_i. $\Delta\omega$ denotes the detuning frequency between qubit transition frequency ϵ and drive frequency ω_1.

Setting the initial condition $\rho_{11}(t=0) = 1$, the diagonal entries in Eq. (2.14) imply that a qubit that was prepared in its excited state $|\mathrm{e}\rangle$ will relax exponentially to its groundstate $|\mathrm{g}\rangle$ at a rate Γ_1.

The dephasing of a qubit is measured in a Ramsey experiment [Ram49]. The qubit dynamics in the equatorial plane is experimentally acquired by rotating its Bloch vector to the equator. At the end of a measurement sequence, the Bloch vector is rotated back to the quantization axis of the qubit prior to state readout, see Sec. 2.5. Both rotations are achieved by applying $\pi/2$ pulses that wind the qubit Bloch vector around a fixed axis in the equatorial plane. By preparing the qubit along the x-axis in a state $|x\rangle = 1/\sqrt{2}(|g\rangle + |e\rangle)$ such that $\hat{\rho}(0) = |x\rangle\langle x|$, we obtain $c_0 = c_1 = 1/2$. From Eq. (2.14), the probability to find the qubit in state $|x\rangle$, corresponding to the probability of finding the qubit in $|e\rangle$ after the final $\pi/2$-rotation, is

$$\langle x|\hat{\rho}(t)|x\rangle = \frac{1}{2} + \frac{1}{2}e^{-\Gamma_2^* t}\cos\Delta\omega t. \tag{2.15}$$

For zero detuning, $\Delta\omega = 0$, Eq. (2.15) describes an exponential damping of the qubit state at the dephasing rate Γ_2^* to a randomized state in the equatorial plane of the Bloch sphere where all phase information is lost. For $\Delta\omega > 0$, we observe oscillations at the detuning frequency $\Delta\omega$, which are referred to as Ramsey fringes.

2.1.3 Rabi oscillations

The coherent oscillations between two states of a qubit in the presence of an external driving field are called Rabi oscillations. It is therefore a first check experiment to verify qubit behaviour and is used to calibrate pulses required for qubit excitation and gate operations.

The Hamiltonian of a driven qubit reads

$$\frac{\hat{H}}{\hbar} = \frac{\epsilon}{2}\hat{\sigma}_z + \Omega_R\hat{\sigma}_x\cos\omega t, \tag{2.16}$$

with drive amplitude Ω_R denoting the Rabi frequency and a drive frequency ω. The drive couples transversally and therefore via Pauli's operator $\hat{\sigma}_x$. The unitary transformation $\hat{U}$ for changing into the frame rotating at the drive frequency ω reads

$$\hat{U} = \exp\left\{i\frac{\omega}{2}\hat{\sigma}_z t\right\}. \tag{2.17}$$

From the definition of a state transformation $|\tilde{\psi}\rangle = \hat{U}|\psi\rangle$ we find the transformed Hamiltonian $\tilde{H}$ according to

$$\tilde{H} = \hat{U}\hat{H}\hat{U}^\dagger - i\hat{U}\dot{\hat{U}}^\dagger. \tag{2.18}$$

This can be directly seen by plugging into the time dependent Schrödinger equation

$$\begin{aligned}
\hat{H}|\psi\rangle &= i\frac{\partial}{\partial t}|\psi\rangle \\
\hat{H}\hat{U}^\dagger|\tilde{\psi}\rangle &= i\frac{\partial}{\partial t}\left(\hat{U}^\dagger|\tilde{\psi}\rangle\right) \\
&= i\dot{\hat{U}}^\dagger|\tilde{\psi}\rangle + \hat{U}^\dagger i\frac{\partial}{\partial t}|\tilde{\psi}\rangle \\
\Rightarrow \hat{U}\hat{H}\hat{U}^\dagger|\tilde{\psi}\rangle - i\hat{U}\dot{\hat{U}}^\dagger|\tilde{\psi}\rangle &= i\frac{\partial}{\partial t}|\tilde{\psi}\rangle = \tilde{H}.
\end{aligned} \tag{2.19}$$

Applying the transformation in Eq. (2.17) to the driven Hamiltonian, Eq. (2.16) yields

$$\frac{\tilde{H}}{\hbar} = \frac{\epsilon - \omega}{2}\hat{\sigma}_z + \frac{\Omega_\mathrm{R}}{2}\hat{\sigma}_x = \frac{\Delta\omega}{2}\hat{\sigma}_z + \frac{\Omega_\mathrm{R}}{2}\hat{\sigma}_x, \tag{2.20}$$

with detuning frequency $\Delta\omega = \epsilon - \omega$. Diagonalization of the Hamiltonian in Eq. (2.20) yields the eigenenergies

$$E_\pm = \pm\frac{1}{2}\sqrt{\Omega_\mathrm{R}^2 + \Delta\omega^2} = \pm\frac{1}{2}\Omega'_\mathrm{R}, \tag{2.21}$$

defining the generalized Rabi frequency $\Omega'_\mathrm{R} = \sqrt{\Omega_\mathrm{R}^2 + \Delta\omega^2}$. The corresponding eigenvectors forming the new basis states become

$$\begin{pmatrix} |+\rangle \\ |-\rangle \end{pmatrix} = \begin{pmatrix} \frac{\Omega'_\mathrm{R} - \Delta\omega}{\sqrt{2\Omega_\mathrm{R}^2 - 2\Omega'_\mathrm{R}\Delta\omega}} & \frac{\Omega_\mathrm{R}}{\sqrt{2\Omega_\mathrm{R}^2 - 2\Omega'_\mathrm{R}\Delta\omega}} \\ \frac{-\Omega'_\mathrm{R} - \Delta\omega}{\sqrt{2\Omega_\mathrm{R}^2 + 2\Omega'_\mathrm{R}\Delta\omega}} & \frac{\Omega_\mathrm{R}}{\sqrt{2\Omega_\mathrm{R}^2 + 2\Omega'_\mathrm{R}\Delta\omega}} \end{pmatrix} \begin{pmatrix} |\mathrm{g}\rangle \\ |\mathrm{e}\rangle \end{pmatrix}. \tag{2.22}$$

The time evolution operator in the $|\pm\rangle$ basis reads

$$\hat{U}(t) = \exp\left\{-\mathrm{i}\frac{\Omega'_\mathrm{R}t}{2}\left(|-\rangle\langle-| - |+\rangle\langle+|\right)\right\} \tag{2.23}$$

and the probability to find the qubit in state $|\mathrm{g}\rangle$ becomes [Sak94]

$$p_{|\mathrm{g}\rangle} = \left|\langle\mathrm{g}|\,\hat{U}(t)\,|\mathrm{g}\rangle\right|^2 \tag{2.24}$$

for initial qubit state $|\mathrm{g}\rangle$. In order to evaluate Eq. (2.24), the original qubit basis state $|\mathrm{g}\rangle$ must be expressed in terms of the new basis states $|+\rangle$, $|-\rangle$. By inverting the matrix in Eq. (2.22), we obtain

$$|\mathrm{g}\rangle = \sqrt{\frac{\Omega'_\mathrm{R} - \Delta\omega}{2\Omega'_\mathrm{R}}}\,|+\rangle - \sqrt{\frac{\Omega'_\mathrm{R} + \Delta\omega}{2\Omega'_\mathrm{R}}}\,|-\rangle. \tag{2.25}$$

Plugging into Eq. (2.24) yields

$$\begin{aligned} p_{|\mathrm{g}\rangle} &= \left|\frac{\Omega'_\mathrm{R} - \Delta\omega}{2\Omega'_\mathrm{R}} e^{\mathrm{i}\Omega'_\mathrm{R}t/2} + \frac{\Omega'_\mathrm{R} + \Delta\omega}{2\Omega'_\mathrm{R}} e^{-\mathrm{i}\Omega'_\mathrm{R}t/2}\right|^2 \\ &= \frac{1}{2}\left(1 + \left(\frac{\Delta\omega}{\Omega'_\mathrm{R}}\right)^2\right) + \frac{1}{2}\left(1 - \left(\frac{\Delta\omega}{\Omega'_\mathrm{R}}\right)^2\right)\cos\Omega'_\mathrm{R}t \end{aligned} \tag{2.26}$$

and accordingly we find $p_{|\mathrm{e}\rangle} = |\langle\mathrm{e}|\,\hat{U}(t)\,|\mathrm{g}\rangle|^2 = 1 - p_{|\mathrm{g}\rangle}$.

For a resonant Rabi drive, $\Delta\omega = 0$, Eq. (2.26) describes harmonic oscillations between the qubit states $|\mathrm{g}\rangle$, $|\mathrm{e}\rangle$ with a Rabi frequency Ω_R, corresponding to the applied drive amplitude. For departing from the resonance condition, $\Delta\omega \neq 0$, we obtain an increased generalized Rabi frequency $\Omega'_{\mathrm{R}} > \Omega_{\mathrm{R}}$. Additionally, the oscillation amplitude is decreased such that the state $|\mathrm{e}\rangle$ is never reached. The reason is an effective tilt of the rotation axis with respect to the Bloch sphere coordinate system, which is reflected by the principle axis transformation during diagonalization of Eq. (2.20).

2.2 Superconductivity

The coherent quantum properties of superconductors render them a valuable basic resource for quantum circuits. Apart from their intriguing microscopic characteristics, superconductors reveal remarkable macroscopic phenomena. The most striking one is the absolute disappearance of electrical resistivity below a critical temperature T_{c}, first observed in 1911 in mercury [Kam11]. When increasing the temperature, superconductors undergo a second-order phase transition at T_{c} to become a normal metal. Another hallmark property of superconductivity is the Meissner-Ochsenfeld effect [MO33]. It denotes the property of superconductors to exclude magnetic fields, similar to perfect conductors, but additionally to expel a magnetic field that was originally penetrating the normal conductor before cooling below T_{c}. Due to this screening field that precisely cancels the external field, superconductors behave as perfect diamagnets [AM76]. In addition, this implies that superconductivity is destroyed at a certain critical magnetic field [Tin04].

2.2.1 Cooper instability and BCS-Theory

The microscopic theory of superconductivity by Bardeen, Cooper and Schrieffer (BCS) captures and explains all basic phenomena of conventional superconductivity [BCS57]. It allows for a quantitative prediction of the relevant macroscopic parameters of superconductors that were phenomenologically found within the Ginzburg-Landau theory [GL50].

The BCS theory requires a net attractive interaction between electrons close to the Fermi surface of a metal. While the bare Coulomb interaction between two isolated electrons is repulsive, it is predominantly the ion cores that move in response to the electronic motion, leading to a screening of the actual charge of the electron as seen from far away [AM76; Tim12; Tin04]. For electron energies smaller than the Debye frequency ω_{D}, which denotes the typical phonon energy scale, this leads to an effective attraction between electrons and to the formation of Cooper pairs at the Fermi surface.

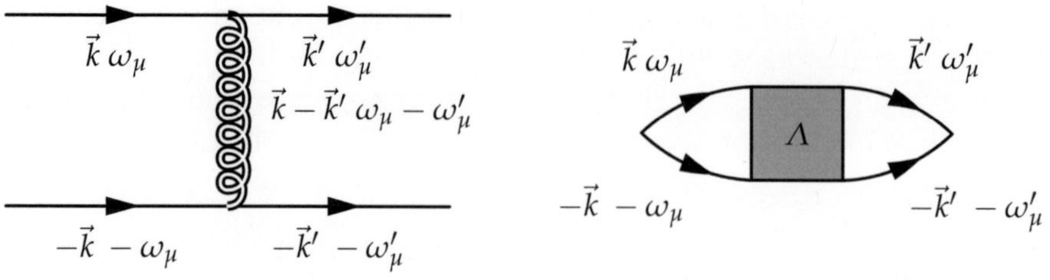

Figure 2.2: Diagrammatic representation of electron-phonon scattering (a) Single scattering event between electrons with momenta $\vec{k}$ and $-\vec{k}$. (b) Cooperon diagram picturing repeated scattering events of a pair of electrons with momenta $\vec{k}$ and $-\vec{k}$. Multiple scattering events are encapsulated in the scattering vertex Λ.

The Green's function formalism provides an elegant method to swiftly arrive at the Cooper instability equation. A single scattering event due to an effective attractive potential

$V_{\text{eff}} =$  (2.27)

is diagrammatically depicted in Fig. 2.2(a).

Multiple scattering events are captured by the Cooperon diagram, see Fig. 2.2(b). The scattering vertex Λ only contains the dominant diagrams without crossings,

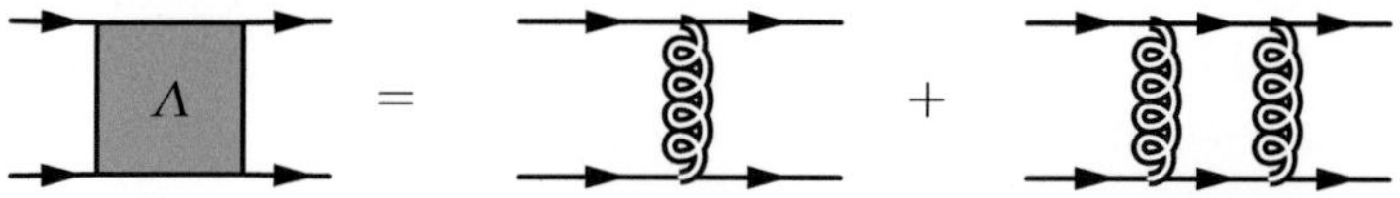

(2.28)

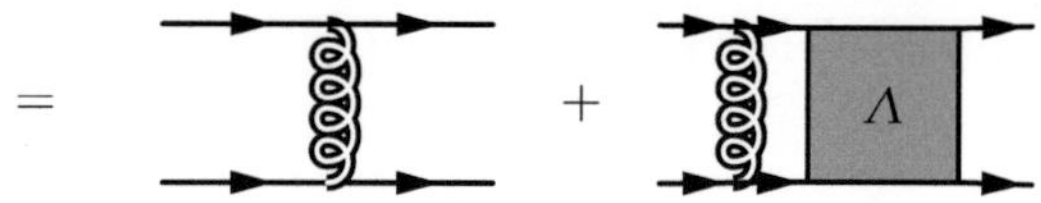

(2.29)

Evaluating Eq. (2.29) with mathematical expressions yields

$$\Lambda = V + \frac{V}{\beta} \sum_{\vec{k},\omega_\mu} G(\vec{k},\omega_\mu) G(-\vec{k},-\omega_\mu)\Lambda \tag{2.30}$$

$$\Rightarrow \Lambda = \frac{V}{1 - \frac{V}{\beta}\sum_{\vec{k},\omega_\mu} G(\vec{k},\omega_\mu) G(-\vec{k},-\omega_\mu)}, \tag{2.31}$$

where we set $V_{\text{eff}} = V$ under the condition that $|\omega_\mu| < \omega_{\text{D}}$ and $V_{\text{eff}} = 0$ otherwise. ω_μ denotes the energy of an unpaired electron relative to the Fermi energy and the sign of V is chosen such that the interaction is attractive.

The scattering instability, known as Cooper instability, manifests in a diverging scattering vertex Λ at $\beta = \beta_\mathrm{c} = (k_\mathrm{B} T_\mathrm{c})^{-1}$, where

$$1 = \frac{V}{\beta_\mathrm{c}} \sum_{\vec{k},\omega_\mu} G(\vec{k},\omega_\mu) G(-\vec{k},-\omega_\mu). \tag{2.32}$$

Using explicit forms of the Green's functions yields the Cooper instability equation

$$1 = \frac{V}{\beta_\mathrm{c}} N(0) \sum_{|\omega_\mu|<\omega_D} \int_{-\infty}^{\infty} \mathrm{d}\varepsilon \frac{1}{\omega_\mu^2 + \varepsilon^2} = \frac{2V}{\beta_\mathrm{c}} N(0) \pi \sum_{n=0}^{\beta\hbar\omega_\mathrm{D}/2\pi} \frac{\beta_\mathrm{c}}{(2n+1)\pi}, \tag{2.33}$$

with $N(0)$ the constant quasi-particle density of states at the Fermi energy. In the last step of Eq. (2.33) we make use of the residue theorem and employ fermionic Matsubara frequencies ω_μ. Assuming $\beta\hbar\omega_\mathrm{D} \gg 1$ and using the Euler constant γ_E yields

$$k_\mathrm{B} T_\mathrm{c} = \frac{2e^{\gamma_E}}{\pi} \hbar\omega_\mathrm{D} \exp\left(-\frac{1}{VN(0)}\right). \tag{2.34}$$

At T_c, the scattering becomes infinitely strong, indicating that the equilibrium state cannot be obtained by perturbation theory applied to the Fermi liquid. This is affirmed mathematically by the lack of an analytic Taylor expansion in the interaction potential V of the Cooper instability equation, Eq. (2.34) [Tim12]. The criticality condition is satisfied regardless of how small the attractive interaction is.

Due to the bosonic nature of a Cooper pair, consisting of two electrons with opposite spin and momentum, each pair can occupy the same quantum mechanical state. The Cooper pairs condensate into a common BCS groundstate, described by a single macroscopic wave function [Ann11]

$$\Psi(\vec{r}) = |\Psi| e^{\mathrm{i}\varphi(\vec{r})}, \tag{2.35}$$

with a collective phase $\varphi(\vec{r})$ dependent on the coordinate $\vec{r}$. Its absolute square $|\Psi(\vec{r})|^2$ is related to the Cooper pair density.

The absolute value $|\Psi|$ of the wave function appears as an order parameter in the Ginzburg-Landau theory, which was later identified with the superconducting gap Δ that opens in the quasi-particle density of states of a superconductor. For a spatial inhomogeneity in the gap parameter $\Delta = \Delta(\vec{r})$, the BCS theory becomes very complicated and the Ginzburg -Landau theory can provide much more reliable predictions [Tin04].

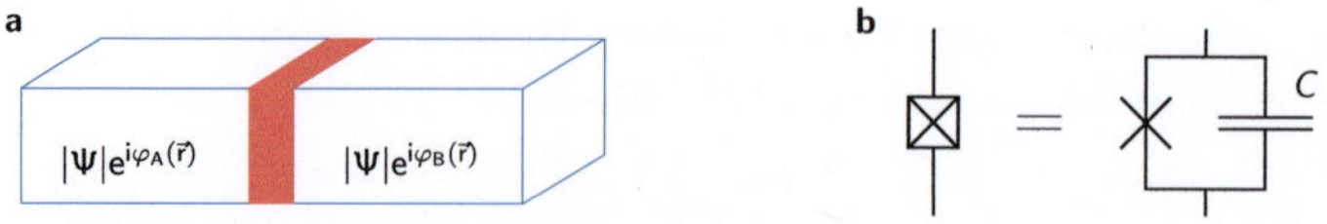

Figure 2.3: Schematic diagram of a Josephson junction (a) Two superconducting electrodes (blue) are separated by a weak link (red). In thermal equilibrium, the amplitudes $|\Psi|$ of the Ginzburg-Landau wave functions of the adjacent superconductors are equal, while their relative phase difference $\Delta\varphi = \varphi_A - \varphi_B$ determines the behaviour of the Josephson junction. (b) Schematic circuit diagram of a Josephson junction. The cross symbolizes the weak link while a square around the cross accounts for the intrinsic capacitance C of the Josephson junction.

2.2.2 Flux quantization

A key phenomenon that is widely used in superconducting quantum devices is magnetic flux quantization in a closed superconducting loop [Sch97]. The second London equation in its local gauge invariant form reads

$$\vec{j} \propto \vec{A} - \frac{\hbar}{2e}\nabla\varphi, \tag{2.36}$$

where $\vec{j}$ denotes the current density and $\vec{A}$ the vector potential. Choosing a closed integration path well within the superconductor, where screening currents vanish, we obtain

$$0 = \oint \vec{A}\cdot \mathrm{d}\vec{r} - \frac{\hbar}{2e}\oint \nabla\varphi \cdot \mathrm{d}\vec{r}. \tag{2.37}$$

From the definition of external flux Φ through a surface S with a contour parametrized by $\mathrm{d}\vec{r}$ and using Stokes' theorem we obtain

$$\Phi = \int \vec{B}\cdot \mathrm{d}\vec{S} = \int \nabla \times \vec{A}\cdot \mathrm{d}\vec{S} = \oint \vec{A}\cdot \mathrm{d}\vec{r} = \frac{\hbar}{2e}\oint \nabla\varphi \cdot \mathrm{d}\vec{r}. \tag{2.38}$$

In order to render the wave function single-valued, the change in phase φ must be an integer multiple of 2π after integrating along the closed contour in Eq. (2.37), such that

$$\Phi = \frac{\hbar}{2e}2\pi n = n\Phi_0,\ n \in \mathbb{Z} \tag{2.39}$$

which defines the magnetic flux quantum $\Phi_0 = h/2e$. The charge $2e$ in the definition of the flux quantum reflects the fact that the supercurrent is carried by pairs of electrons.

2.3 The Josephson effect

The Josephson tunnel junction is the key building block for non-linear superconducting quantum circuits. It consists of two superconducting electrodes that are separated by a weak link [Sch97; Tin04]. The wave functions of the adjacent superconductors overlap and interfere within the barrier region, leading to phase coherent effects determined by the relative phase of the two wave functions. In experiments, a weak link is typically formed by a thin insulating oxide barrier, see Fig. 2.3(a).

As depicted in Fig. 2.3(b), the Josephson junction can be regarded as an ideal tunnel element connected in parallel to a capacitor, accounting for the intrinsic capacitance C of the superconducting electrodes. Tunneling of Cooper pairs through the Josephson junction is described by the Hamiltonian [Dev97]

$$\hat{H}_{\mathrm{J}} = -\frac{E_{\mathrm{J}}}{2} \sum_{N=-\infty}^{\infty} |N\rangle \langle N+1| + |N+1\rangle \langle N| \tag{2.40}$$

with the Josephson energy E_{J}, denoting a macroscopic parameter characteristic for the tunneling rate of a Josephson junction. N counts the number of Cooper pairs on one of the electrodes of the Josephson junction, which allows us to define a charge operator

$$\hat{Q} = -2e\hat{N} \tag{2.41}$$

and a Cooper pair number operator

$$\hat{N} = \sum_{N} N |N\rangle \langle N| \, . \tag{2.42}$$

The tunneling Hamiltonian $\hat{H}_{\mathrm{J}}$ in Eq. (2.40) couples states that differ by one Cooper pair that has tunnelled across the barrier. By defining a phase operator $\hat{\phi}$ as the conjugate of the Cooper pair number operator $\hat{N}$,

$$\left[\hat{\phi}, \hat{N}\right] = \mathrm{i}, \tag{2.43}$$

and corresponding basis states

$$|\phi\rangle = \sum_{N=\infty}^{\infty} e^{\mathrm{i}N\phi} |N\rangle \, , \tag{2.44}$$

we can express the state $|N\rangle$ via Fourier transformation as

$$|N\rangle = \frac{1}{2\pi} \int_0^{2\pi} \mathrm{d}\phi e^{-\mathrm{i}N\phi} |\phi\rangle \, . \tag{2.45}$$

Plugging into Eq. (2.40) yields

$$\hat{H}_{\mathrm{J}} = -\frac{E_{\mathrm{J}}}{2} \frac{1}{2\pi} \int_0^{2\pi} \mathrm{d}\phi \left(e^{\mathrm{i}\phi} + e^{-\mathrm{i}\phi}\right) |\phi\rangle \langle \phi| \, . \tag{2.46}$$

The periodicity of ϕ motivates to introduce an operator [Dev97]

$$e^{i\hat{\phi}} = \frac{1}{2\pi}\int_0^{2\pi} d\phi e^{i\phi} \left|\phi\right\rangle \left\langle\phi\right| \tag{2.47}$$

and Eq. (2.46) becomes

$$\hat{H}_J = -E_J \cos\hat{\phi}. \tag{2.48}$$

The capacitive energy term accounting for the intrinsic Josephson junction capacitance C can be written as

$$\hat{H}_C = \frac{\hat{Q}^2}{2C} = \frac{4e^2}{2C}\hat{N}^2 = 4E_C\hat{N}^2, \tag{2.49}$$

where we introduced the charging energy $E_C = e^2/2C$. This results in a total Hamiltonian

$$\hat{H}_{JJ} = H_C + H_J = 4E_C\hat{N}^2 - E_J\cos\hat{\phi} \tag{2.50}$$

for the Josephson junction.

We now can extract both Josephson relations by evaluating

$$\hat{I} = \frac{d}{dt}\hat{Q} = -\frac{2e}{i\hbar}\left[\hat{N},\hat{H}_{JJ}\right] = -\frac{2e}{\hbar}\frac{\partial\hat{H}_{JJ}}{\partial\hat{\phi}} = \frac{2e}{\hbar}E_J\sin\hat{\phi} = I_c\sin\hat{\phi} \tag{2.51}$$

$$\frac{\partial}{\partial t}\hat{\phi} = \frac{1}{i\hbar}\left[\hat{\phi},\hat{H}_{JJ}\right] = -\frac{1}{\hbar}\frac{\partial\hat{H}_{JJ}}{\partial\hat{N}} = \frac{2e}{\hbar}\hat{U} = \frac{\hat{Q}}{C} = \frac{2e}{\hbar}\hat{V}, \tag{2.52}$$

where we defined the critical current $I_c = 2eE_J/\hbar$ of the Josephson junction. I_c depends on the superconducting properties of its electrodes as well as on the sheet resistance of the barrier. The phase ϕ in the above equations is the gauge-invariant equivalent to the phase difference $\Delta\varphi = \varphi_A - \varphi_B$ of the wave functions of the two adjacent superconductors forming the Josephson junction, see Fig. 2.3(a).

It was the remarkable prediction by Josephson [Jos62] that a zero-voltage supercurrent of Cooper pairs can be sustained below the critical current I_c, before the Josephson junction becomes resistive, see Eq. (2.51). In addition, the phase difference ϕ will evolve in time according to Eq. (2.52) when a voltage V is applied across the Josephson junction.

The hysteretic current-voltage characteristic of a Josephson junction can be understood within the Resistively and Capacitively Shunted Junction (RCSJ) model. It describes a parallel connection of an ideal Josephson junction, a resistance R and a capacitance C. The RCSJ model can explain the non-vanishing voltage across a Josephson junction after reducing the drive current below I_c by the finite inertia of a virtual phase particle travelling down the 'tilted washboard potential' [Tin04]. In addition, the model yields a plasma frequency

$$\omega_\mathrm{p} = \sqrt{\frac{2\pi I_\mathrm{c}}{\Phi_0 C}} \tag{2.53}$$

of the Josephson junction, which yields a non-linear Josephson inductance

$$L_\mathrm{J} = \frac{\Phi_0}{2\pi I_\mathrm{c} \cos\phi}. \tag{2.54}$$

This non-linear current-phase relation is the basis for the anharmonicity in superconducting circuits and allows for the definition of a computational qubit basis [CW08; SG08]. The damping of a Josephson junction is quantified by the Stewart-McCumber parameter β_c [McC68; Ste68], defined as the square of the quality factor Q of a parallel LCR resonator,

$$\beta_\mathrm{c}^{1/2} = Q = \omega_\mathrm{p} RC. \tag{2.55}$$

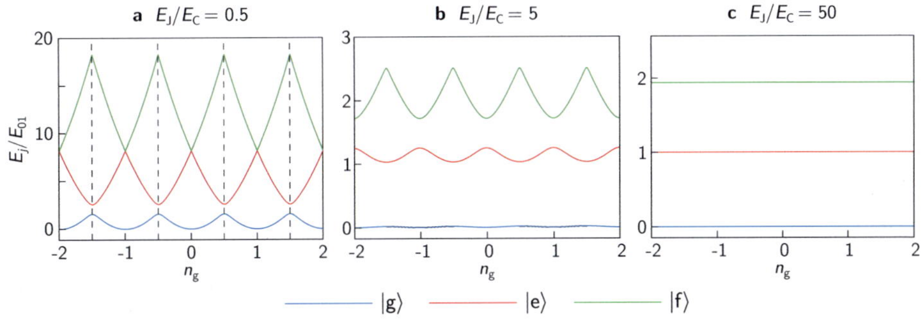

Figure 2.4: Qubit energy dispersions for various ratios of $E_\mathrm{J}/E_\mathrm{C}$ The plots show the eigenenergies E_j of the Hamiltonian, Eq. (2.56), normalized to the fundamental transition energy E_{01} and as a function of the offset charge n_g. E_{01} is evaluated at half integer n_g and an energy offset is subtracted in each plot. The coloured energy bands correspond to the first three energy levels in the harmonic approximation, see Eq. (2.61), denoted as $|\mathrm{g}\rangle$, $|\mathrm{e}\rangle$ and $|\mathrm{f}\rangle$ for $j = 1, 2, 3$, respectively. (a) Energy dispersion for the Cooper pair box regime. The parabolas from the electrostatic energy term can still be identified while the degeneracy at half integer n_g (dashed lines) is lifted by the Josephson coupling term. (b) Intermediate regime, where E_n is well approximated by a tight-binding treatment [Koc+07]. (c) Transmon regime with $E_\mathrm{J}/E_\mathrm{C} \gtrsim 50$ revealing and flat energy levels. As can be seen from the scale, the anharmonicity for the transmon qubit is negative. The plots are similarly given in Ref. [Koc+07].

2.4 The transmon qubit

The qubit of choice for this thesis is the transmon qubit. Its architecture is closely related to the one of the single Cooper pair box (CPB) [Bou+98; NPT99]. The CPB consists of a superconducting island that is connected via a Josephson junction to a large superconducting reservoir. The gate potential and thereby the offset charge n_g on the island is controlled via a gate electrode that capacitively couples to the island. The CPB operates in a regime where $E_J \ll E_C$. Assuming an intrinsic Josephson junction capacitance $C = 1\,\mathrm{fF}$ and an experimental temperature $T = 25\,\mathrm{mK}$, the charging energy $E_C = e^2/2C = 80\,\mathrm{\mu eV}$ required to add a single electron on the CPB island is much bigger than the thermal energy $k_B T \sim 2\,\mathrm{\mu eV}$ [CW08].

Up to an additional term accounting for the offset charge n_g, the Hamiltonian of the CPB is identical to the one of a single Josephson junction, Eq. (2.50),

$$\hat{H} = 4E_C(\hat{N} - n_g)^2 - E_J \cos\hat{\phi}. \tag{2.56}$$

In analogy to the definition in Sec. 2.3, $\hat{N}$ counts the number of excess Cooper pairs on the island. The eigenenergies of the Cooper pair box are depicted in Fig. 2.4(a). The state of the CPB is encoded in $\hat{N}$, corresponding to different parabolas visible. It is operated at half integer n_g, where the energy degeneracy is lifted by the weak Josephson coupling term and a coherent superposition of different charge states is present [NPT99].

The transmon qubit was proposed in Ref. [Koc+07] as a 'transmission line shunted plasma oscillation qubit' and is since then the most promising type of qubit employed to advance scalable quantum computation based on superconducting circuits. It differs from the CPB by an additional large shunt capacitance in parallel to the Josephson junction and is therefore described by the identical Hamiltonian, Eq. (2.56), with a decreased charging energy $E_C \propto C^{-1}$. Due to the strongly increased ratio of $E_J/E_C \gtrsim 50$, the Josephson term is dominant, which causes the transmon to operate in the phase regime. Here, the charge is not well-defined, according to the commutation relation provided in Eq. (2.43). The key advantage of the transmon qubit is its exponentially decreased sensitivity to charge noise, as apparent in Fig. 2.4 by a flattening of the energy dispersion for increasing the ratio E_J/E_C. This comes at the expense of an algebraic decrease in qubit anharmonicity

$$\alpha = E_{12} - E_{01} \tag{2.57}$$

where $E_{ij} = E_j - E_i$ and E_j denoting the transmon eigenenergies as visible in Fig. 2.4(c). Figure 2.4(b) shows the dispersion for an intermediate regime, while a complete suppression of substructure is found in Fig. 2.4(c), which suggests that a biasing of the transmon is no longer required since the charge has no meaning in the phase regime.

In fact, the transmon is a multi-level quantum circuit that operates in a regime that is very similar to the simple LC-oscillator, with the inductance replaced by a Josephson junction. The non-linearity of the Josephson junction renders the potential energy of the transmon weakly anharmonic, which enables to address individual transitions, see Fig. 2.5(a).

The close resemblance of the transmon qubit to a slightly anharmonic resonator becomes apparent in the harmonic approximation of the transmon Hamiltonian, Eq. (2.56). To this end we expand the potential energy term $-E_\mathrm{J}\cos\hat{\phi}$ up to fourth order in $\hat{\phi}$, yielding

$$-E_\mathrm{J}\cos\hat{\phi} \approx \frac{E_\mathrm{J}}{2}\hat{\phi}^2 - \frac{E_\mathrm{J}}{4!}\hat{\phi}^4 + \text{const.} \tag{2.58}$$

In order to cast Eq. (2.56) into the standard form of the harmonic oscillator for terms of order $O(\hat{\phi}^2)$, we identify $4E_\mathrm{C} \equiv \frac{\hbar^2}{2m}$ and $E_\mathrm{J}/2 \equiv \frac{1}{2}m\omega^2$, where m is the mass of the particle described by the quantum harmonic oscillator at angular frequency ω. This allows us to define the conjugate operators satisfying Eq. (2.43)

$$\hat{\phi} = \left(\frac{2E_\mathrm{C}}{E_\mathrm{J}}\right)^{1/4}(\hat{a}^\dagger + \hat{a}), \quad \hat{N} = \mathrm{i}\left(\frac{E_\mathrm{J}}{32E_\mathrm{C}}\right)^{1/4}(\hat{a}^\dagger - \hat{a}) \tag{2.59}$$

in the representation of bosonic creation (annihilation) operators $\hat{a}^\dagger$ ($\hat{a}$) [Koc+07]. The expansion in Eq. (2.58) is justified since $\hat{\phi}$ is small in the transmon regime where $E_\mathrm{J} \gg E_\mathrm{C}$. Plugging into Eq. (2.56) and taking into account the bosonic commutation relation $[\hat{a}, \hat{a}^\dagger] = 1$ yields

$$\hat{H}_\mathrm{p} = \sqrt{8E_\mathrm{C}E_\mathrm{J}}\hat{a}^\dagger\hat{a} - \frac{E_\mathrm{C}}{2}\left((\hat{a}^\dagger\hat{a})^2 + \hat{a}^\dagger\hat{a}\right) + \text{const.} \tag{2.60}$$

and we can find the transmon eigenenergies E_j to be

$$E_j = \sqrt{8E_\mathrm{C}E_\mathrm{J}}j - \frac{E_\mathrm{C}}{2}(j^2 + j) + \text{const.} \tag{2.61}$$

The transmon excitation number operator $\hat{a}^\dagger\hat{a} = \sum_j j\,|j\rangle\,\langle j|$ in the harmonic approximation should be distinguished from the Cooper pair number operator $\hat{N}$. From Eq. (2.61), we obtain the approximate fundamental transmon transition frequency $\epsilon_{01} = E_{01}/\hbar = \sqrt{8E_\mathrm{C}E_\mathrm{J}}/\hbar - E_\mathrm{C}/\hbar$ and the anharmonicity $\alpha \sim -E_\mathrm{C}$.

A typical relative anharmonicity of $\alpha_\mathrm{r} = \alpha/E_{01} \sim 0.05$ suppresses unwanted excitations of higher transmon levels and thereby allows one to use the transmon as a two-level system by defining a computational qubit basis using the two lowest energy levels, denoted as $|\mathrm{g}\rangle$, $|\mathrm{e}\rangle$ in Fig. 2.4.

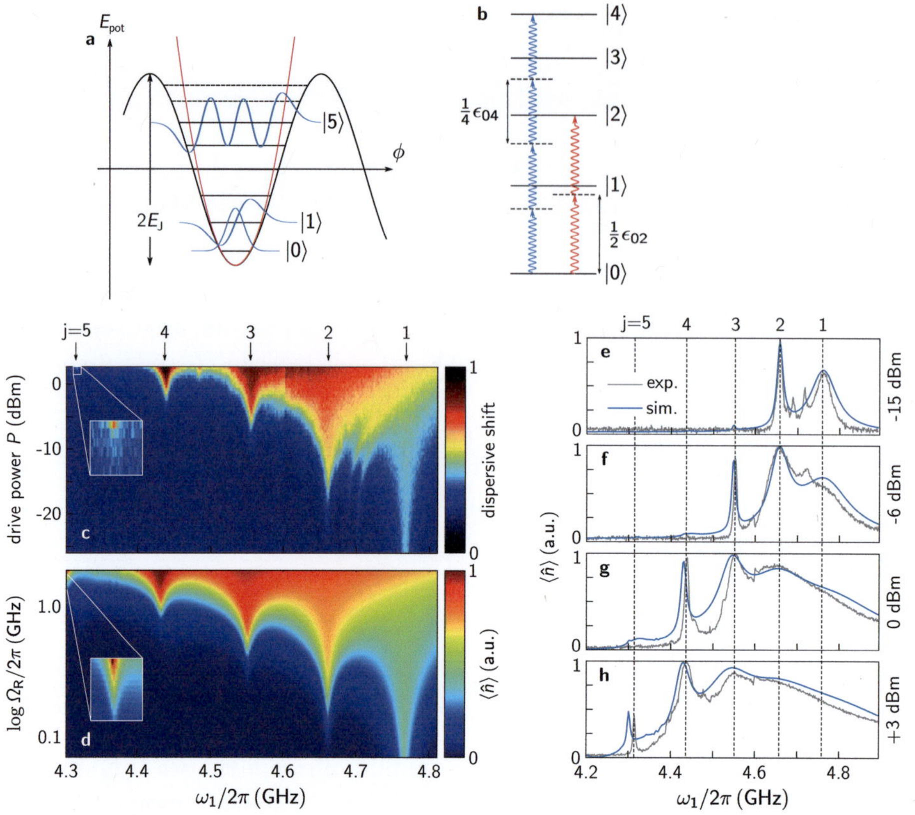

Figure 2.5: Multiphoton transitions in the weakly anharmonic transmon circuit (a) Josephson potential E_{pot} from the Hamiltonian in Eq. (2.56) with indicated approximation as a harmonic potential (red). Transmon levels only up to level $|5\rangle$ are properly bound in the Josephson potential and were experimentally observed. Harmonic oscillator wave functions are schematically indicated in blue. (b) Energy diagram illustrating the observed multiphoton transitions. (c) Spectroscopically measured power spectrum for multiphoton transition with frequency ϵ_{0j}/j. We plot the transmission magnitude of a dispersively coupled readout resonator in colours with respect to an applied drive of frequency ω_1 and the drive power P. (d) Respective master equation simulation with the Rabi frequency Ω_{R} on the vertical axis, corresponding to the drive amplitude. The expectation value $\langle\hat{n}\rangle = \langle\hat{a}^\dagger\hat{a}\rangle$ of the qubit population is plotted in colours. (e)-(h) Detailed comparison of experimental and simulated data for horizontal cuts of the power spectrum. Please refer to Sec. 4 and Ref. [Bra+15] for experimental details.

The exact eigenenergies of the Hamiltonian in Eq. (2.56) can be found by writing the corresponding Schrödinger equation in the phase basis and casting it in the form of Mathieu's differential equation [Ast64; Koc+07; Bra13]. The Schrödinger equation reads

$$\left[4E_{\text{C}}\left(-\mathrm{i}\frac{\mathrm{d}}{\mathrm{d}\phi}-n_{\text{g}}\right)^2 - E_{\text{J}}\cos\phi\right]\psi(\phi) = E\psi(\phi) \tag{2.62}$$

which is equivalent to

$$\frac{\mathrm{d}^2}{\mathrm{d}x^2}g(x) + \left(\frac{E}{E_\mathrm{C}} + \frac{E_\mathrm{J}}{E_\mathrm{C}}\cos(2x)\right)g(x) = 0 \tag{2.63}$$

with $\phi = 2x$ and introducing $g(x) = e^{-2in_g x}\psi(2x)$. The eigenenergies E_j for transmon levels j can be expressed at half integer flux in terms of Mathieu's characteristic values according to

$$E_j = E_\mathrm{C}\left\{\begin{array}{ll} M_\mathrm{A}\left(k+1, -\frac{E_\mathrm{J}}{2E_\mathrm{C}}\right) & \text{for } k = 2j \\ M_\mathrm{B}\left(k, -\frac{E_\mathrm{J}}{2E_\mathrm{C}}\right) & \text{for } k = 2j+1 \end{array}\right\}, j = 1, 2, 3, ... \tag{2.64}$$

M_A and M_B denote the even and odd Mathieu characteristic values.

By evaluating Eq. (2.64) for an experimentally investigated transmon qubit of $E_\mathrm{J}/E_\mathrm{C} \sim 80$, we find the existence of five bound states in the Josephson potential [Bra+15], see Fig. 2.5(a). A numerical diagonalization in the Cooper pair number basis $|N\rangle$ yields a groundstate energy of $E_0/h = 2.4\,\mathrm{GHz}$ which equals the vacuum energy $\hbar\epsilon_{01}/2$ with fundamental transition frequency $\epsilon_{01}/2\pi = 4.78\,\mathrm{GHz}$. Due to the small anharmonicity $\alpha_\mathrm{r} \sim 0.045$ in the transmon circuit, multiphoton transitions of frequency ϵ_{0j}/j can be observed up to $j = 5$ in spectroscopy [Bra+15]. See Fig. 2.5(b) for a schematic illustration and Fig. 2.5(c-h) for a comparison between experimental measurement and master equation simulation.

2.5 Cavity quantum electrodynamics

The field of cavity quantum electrodynamics (cQED) studies the fundamental interaction between matter and light at the archetype system of a quantized harmonic oscillator mode coupled to a two-level atom. The two quantum states of the atom are described via a spin-1/2 degree of freedom, recovering the Bloch sphere description introduced in Sec. 2.1.1 for a qubit.

With the emergence of superconducting circuits as a versatile platform for cQED experiments, the term circuit QED was introduced, standing for a circuit implementation of cQED. It typically involves a transmission line resonator of low internal loss rate that couples to a superconducting qubit [Bla+04; Wal+04].

2.5.1 Jaynes-Cummings Hamiltonian

The coupling in cQED systems is induced by a dipole interaction of the qubit with the oscillating field of a cavity or a circuit resonator. For a capacitive coupling of a transmon qubit, having a dipole moment $\hat{\vec{d}} \propto \vec{d}_0\hat{\sigma}_x$, to the electric field $\hat{\vec{E}} \propto \hat{b}^\dagger + \hat{b}$ of a quantum harmonic oscillator, we find a coupling term

$$\hat{H}_c = \hbar g \hat{\sigma}_x \left(\hat{b}^\dagger + \hat{b}\right) \propto \hat{\vec{E}} \cdot \hat{\vec{d}} \tag{2.65}$$

with coupling strength g. The complete quantum Rabi Hamiltonian [Rab37; Rab36] describing a generic cQED system therefore reads

$$\frac{\hat{H}}{\hbar} = \frac{\epsilon}{2}\hat{\sigma}_z + \omega\hat{b}^\dagger\hat{b} + g\hat{\sigma}_x\left(\hat{b}^\dagger + \hat{b}\right), \tag{2.66}$$

with ϵ the qubit energy splitting and ω the resonator frequency. $\hat{\sigma}_i$ are Pauli matrices and $\hat{b}^\dagger$ ($\hat{b}$) are creation (annihilation) operators in the Fock space of the bosonic oscillator mode, such that $[\hat{b}, \hat{b}^\dagger] = 1$. The described type of interaction is often referred to as transversal coupling, since it appears on the off-diagonals of both the qubit and the resonator Hamiltonians.

In many applications of quantum optics and cQED, the system described by Eq. (2.66) is in a regime where $g \ll \epsilon, \omega$. This enables to apply a rotating wave approximation (RWA), which reduces the quantum Rabi Hamiltonian in Eq. (2.66) to the Jaynes-Cummings Hamiltonian $\hat{H}_{JC}$ [JC63]. In the approximation, so called counter-rotating terms of the form $\hat{\sigma}_+\hat{a}^\dagger + \hat{\sigma}_-\hat{a}$ are neglected, such that

$$\frac{\hat{H}_{JC}}{\hbar} = \frac{\epsilon}{2}\hat{\sigma}_z + \omega\hat{b}^\dagger\hat{b} + g\left(\hat{\sigma}_+\hat{b} + \hat{\sigma}_-\hat{b}^\dagger\right), \tag{2.67}$$

with $\hat{\sigma}_\pm = 1/2\left(\hat{\sigma}_x \pm \mathrm{i}\hat{\sigma}_y\right)$. Applying the RWA renders the resulting Jaynes-Cummings model exactly solvable since it can be expressed in terms of infinitely many equivalent and disconnected subspaces that allow for trivial diagonalization individually [Bra11; HR06]. At the same time, this motivates the investigation of the exact quantum Rabi model in the framework of quantum simulations, revealing non-classical and novel physics, see Sec. 6.

In order to verify the RWA, we follow the formal text book approach of time-independent perturbation theory [Sak94]. The diagonal Hamiltonian H_0 of the problem with trivial solution is given by the subsystem energy terms

$$\frac{H_0}{\hbar} = \frac{\epsilon}{2}\hat{\sigma}_z + \omega\hat{a}^\dagger\hat{a}. \tag{2.68}$$

We consider the counter rotating terms of the coupling term in Eq. (2.66) as a perturbation,

$$\frac{\hat{H}_1}{\hbar} = g\left(\hat{\sigma}_+\hat{a}^\dagger + \hat{\sigma}_-\hat{a}\right). \tag{2.69}$$

Since H_1 couples unperturbed eigenstates of non-degenerate eigenenergies, we can apply non-degenerate perturbation theory. This yields an energy correction

$$\Delta E_{n,\mathrm{e}} = \hbar^2 g^2 \sum_{m,s} \frac{|\langle n,\mathrm{e}|\,\hat{\sigma}_+ \hat{a}^\dagger + \hat{\sigma}_- \hat{a}\,|m,s\rangle|^2}{E_{n,\mathrm{e}} - E_{m,s}} \tag{2.70}$$

$$= \frac{\hbar^2 g^2}{E_{n,\mathrm{e}} - E_{n-1,\mathrm{g}}} = \hbar \frac{g^2}{\epsilon + \omega} \tag{2.71}$$

and analogously $\Delta E_{n,\mathrm{g}} = -\hbar g^2/(\epsilon + \omega)$. The RWA is therefore valid in the regime where $g/(\epsilon + \omega) \ll 1$, which is true in typical scenarios of cavity and circuit QED.

The Jaynes-Cummings Hamiltonian reveals its most remarkable features and becomes most relevant in quantum optics and various cQED platforms in the strong coupling limit, where the energy decay rates of the subsystems are smaller than the mutual coupling strength. It is one of the major benefits of circuit QED that the strong coupling condition can straightforwardly be achieved [Wal+04].

Since the interaction term of the Jaynes-Cummings Hamiltonian in Eq. (2.67) conserves the total excitation number and couples only neighbouring qubit and resonator levels, the Hamiltonian takes a block diagonal shape in matrix form, consisting of a set of disconnected two-dimensional sub-matrices for any resonator excitation number n [Sch07]. This allows for an exact diagonalization of the n-th sub-matrix

$$\frac{\hat{H}_n}{\hbar} = \begin{pmatrix} (n+1)\omega & g\sqrt{n+1} \\ g\sqrt{n+1} & n\omega + \epsilon \end{pmatrix}. \tag{2.72}$$

Its eigenenergies are

$$\frac{E_{n,\pm}(\Delta)}{\hbar} = \omega\left(n + \frac{1}{2}\right) + \frac{\epsilon}{2} \pm \frac{1}{2}\sqrt{4g^2(n+1) + \Delta^2} \tag{2.73}$$

with a mutual detuning $\Delta = \epsilon - \omega$.

The significance of strong coupling in the Jaynes-Cummings model becomes most apparent in the case where the transition frequencies of atom and resonator are degenerate. Equation (2.73) reduces to

$$\frac{E_{n,\pm}(\Delta = 0)}{\hbar} = \omega(n+1) \pm g\sqrt{n+1}, \tag{2.74}$$

illustrating that the frequencies of the new eigenstates are split by $2g\sqrt{n+1}$ [Bla+04]. An analogy to a classical treatment, the levels of the new eigenstates repel each other and are clearly distinguishable when the strong coupling condition is met. The frequency splitting in the single-photon regime therefore becomes $2g$, which is a property that is commonly exploited for characterization purposes in qubit spectroscopy experiments.

The eigenstates of the coupled system are no longer of tensor product form $|n, \mathrm{g}\rangle$, $|n, \mathrm{e}\rangle$, but are superposition states of the form

$$|n, \pm\rangle = \frac{1}{\sqrt{2}} \left(|n+1, \mathrm{g}\rangle \pm |n, \mathrm{e}\rangle\right). \tag{2.75}$$

2.5.2 Dispersive limit

The regime where the detuning between qubit and resonator is large compared to the coupling strength, $g/\Delta \ll 1$, is called the dispersive limit [Bla+04]. In this case, qubit and resonator maintain their individual character while the coupling is merely a perturbation that leads to a dressing of the respective energy levels. In the dispersive limit, the Jaynes-Cummings Hamiltonian in Eq. (2.67) can be diagonalized approximately by applying the canonical transformation

$$\hat{U} = \exp\left\{\frac{g}{\Delta}\left(\hat{\sigma}_+ \hat{a} - \hat{\sigma}_- \hat{a}^\dagger\right)\right\}. \tag{2.76}$$

Expanding up to order $(g/\Delta)^2$ yields [Bla+04]

$$\frac{\hat{U}\hat{H}_{\mathrm{JC}}\hat{U}^\dagger}{\hbar} \approx \left(\omega + \frac{g^2}{\Delta}\hat{\sigma}_z\right)\hat{a}^\dagger \hat{a} + \frac{1}{2}\left(\epsilon + \frac{g^2}{\Delta}\right)\hat{\sigma}_z \tag{2.77}$$

$$\approx \left(\omega + \chi\hat{\sigma}_z\right)\hat{a}^\dagger \hat{a} + \frac{1}{2}\left(\epsilon + \chi\right)\hat{\sigma}_z. \tag{2.78}$$

Hence, the dressing in the dispersive regime leads to a shift of the qubit and resonator frequencies by the dispersive shift $\chi = g^2/\Delta$. In particular, the effective resonator frequency $\omega \pm \chi$ is dependent on the qubit state, which enables perform a projective quantum non-demolition measurement of the qubit state along its quantization axis. When populating the resonator, that acts as the qubit readout device with photons, the qubit Bloch vector collapses into one of its basis states $|\mathrm{g}\rangle$, $|\mathrm{e}\rangle$. Since the qubit state and the measured outcome are equal, the measurement is referred to as non-demolition [Sch07]. This is an essential feature of the dispersive readout scheme, enabling complex gate sequences conditional on a measured qubit state by using a fast quantum feedback.

The dispersive shift obtained in Eq. (2.77) is in agreement with the exact result in Eq. (2.73) which takes the form

$$\frac{E_{n=0,\pm}}{\hbar} = \frac{\omega + \epsilon}{2} \pm \frac{\Delta}{2}\sqrt{1 + \frac{4g^2}{\Delta^2}} \approx \frac{\omega + \epsilon}{2} \pm \left(\frac{\Delta}{2} + \frac{g^2}{\Delta}\right) \tag{2.79}$$

in the limit where $g/\Delta \ll 1$. Equation (2.79) yields the dispersively shifted energies of qubit and resonator as

$$E_+/\hbar = \epsilon + \chi \quad E_-/\hbar = \omega - \chi. \tag{2.80}$$

When using the transmon as an effective two-level qubit, we typically plot the qubit population on a scale from 0 to 1, where 0 corresponds to the dispersive resonator shift when the qubit is in $|g\rangle$ and 1 corresponds to the dispersive shift when the qubit is in $|e\rangle$. The qubit population is therefore a probability for finding the qubit in $|e\rangle$, based on the previously calibrated qubit basis.

The presented concepts remain valid when treating the transmon qubit as a many-level circuit [Koc+07; Nee+09]. The approximately diagonalized Hamiltonian, Eq. (2.78) takes the form

$$\begin{aligned}\frac{\hat{H}'}{\hbar} =& \sum_j \omega_j |j\rangle\langle j| + \sum_{j=1} \chi_{j-1,j} |j\rangle\langle j| \\ &+ \hat{a}^\dagger \hat{a} \left(\omega - \chi_{01} |0\rangle\langle 0| + \sum_{j=1} \left(\chi_{j-1,j} - \chi_{j,j+1}\right) |j\rangle\langle j| \right),\end{aligned} \tag{2.81}$$

where ω_j are the eigenenergies of the transmon qubit and the dispersive shift $\chi_{j,j+1} = g^2_{j,j+1}/(\epsilon_{j,j+1} - \omega)$ induced by transitions between neighbouring transmon levels $j, j+1$ with $\epsilon_{j,j+1} = \epsilon_{j+1} - \epsilon_j$ and $\epsilon_j = E_j/\hbar$. We experimentally analysed the back-action of higher order multiphoton transitions on the readout resonator in a regular Jaynes-Cummings system and demonstrated quantitative accordance with the theoretical expectation from Eq. (2.81). With a pump-probe setup in addition to the readout tone, we could achieve a multiphoton dressing of the transmon circuit and were able to observe the emerging Rabi-sidebands and Autler-Townes like splittings involving up to five levels of the circuit. These results are published in Ref. [Bra+15].

3 Microwave theory

This chapter provides the fundamentals of microwave theory necessary for evaluating and characterising quantum circuits. I give a brief summary of transmission line theory and its circuit implementation which is followed by some network analysis techniques. The description of microwave resonators is carried out in more detail. Despite of its importance for a large community of microwave engineers as well as to the field of circuit based quantum information processing, I find a lack of a comprehensive treatment of respective fitting models in literature. To this end, I derive relevant formulas relating the experimentally accessible scattering matrix elements with the frequency response of microwave resonators, providing the mathematical framework for extracting quality factors of resonating networks based on circuit impedances.

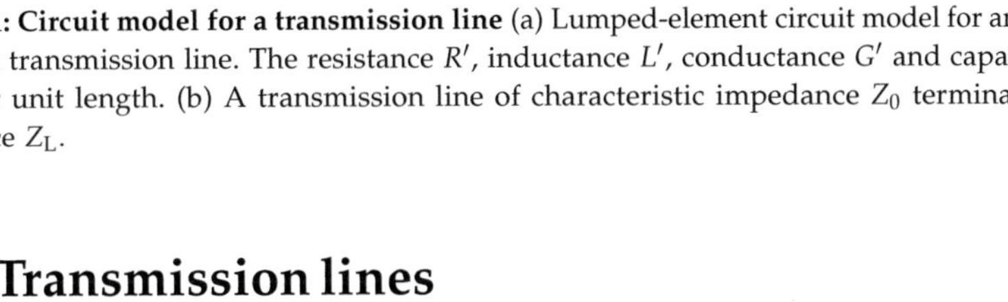

Figure 3.1: Circuit model for a transmission line (a) Lumped-element circuit model for an infinitesimal piece of a transmission line. The resistance R', inductance L', conductance G' and capacitance C' are given per unit length. (b) A transmission line of characteristic impedance Z_0 terminated in a load impedance Z_L.

3.1 Transmission lines

In contrast to dc electronics, where a conducting wire equalizes the respective potentials at its two ends, a transmission line must be considered as a distributed element network. This is caused by the fact that the size of a transmission line can be similar to the wavelength of the propagating wave, such that spatial inhomogeneities in voltage and current need to be considered [Poz12]. An equivalent circuit for a generic transmission line is depicted in Fig. 3.1(a). Kirchhoff's rules yield the telegrapher equations [Poz12]

$$\frac{\mathrm{d}V(z)}{\mathrm{d}z} = -\,(R' + \mathrm{i}\omega L')I(z) \tag{3.1}$$

$$\frac{\mathrm{d}I(z)}{\mathrm{d}z} = -\,(G' + \mathrm{i}\omega C')V(z), \tag{3.2}$$

with resistance R', inductance L', conductance G' and capacitance C' given per unit length. Equations (3.1), (3.2) are solved by a plane wave ansatz, yielding

$$V(z) = V_0^+ e^{-\gamma z} + V_0^- e^{\gamma z} \tag{3.3}$$

$$I(z) = \frac{V_0^+}{Z_0} e^{-\gamma z} - \frac{V_0^-}{Z_0} e^{\gamma z}. \tag{3.4}$$

V_0^+ (V_0^-) denote the voltage of the wave travelling to the right (left) and we define a propagation constant $\gamma = \sqrt{(R' + \mathrm{i}\omega L')(G' + \mathrm{i}\omega C')}$. In Eq. (3.4) we introduced the characteristic impedance Z_0 of the transmission line, which can be found as

$$Z_0 = \sqrt{\frac{R + \mathrm{i}\omega L}{G + \mathrm{i}\omega C}} \tag{3.5}$$

by using Eq. (3.1), (3.2) and Eq. (3.4).

Terminating a transmission line of characteristic impedance Z_0 in a load impedance Z_L amounts to wave reflection dependent on the boundary condition given by Z_L, see Fig. 3.1(b). The total voltage $V(z)$ on a transmission line at position z is given by the sum of right and left travelling waves and the total current $I(z)$ flowing in the positive direction is the difference between current flowing to the right and current flowing to the left. Evaluating Eq. (3.3), (3.4) at $z = 0$ yields

$$Z_\mathrm{L} = \frac{V(z=0)}{I(z=0)} = \frac{V_0^+ + V_0^-}{V_0^+ - V_0^-} Z_0 \tag{3.6}$$

and we obtain the voltage reflection coefficient

$$\Gamma \equiv \frac{V_0^-}{V_0^+} = \frac{Z_\mathrm{L} - Z_0}{Z_\mathrm{L} + Z_0}. \tag{3.7}$$

It is apparent from Eq. (3.7) that the input impedance of a transmission line terminated with a load $Z_\mathrm{L} = Z_0$ does not cause any reflections.

The superposition of incident and reflected waves leads to a spatially varying line impedance [Poz12]. Using Eq. (3.3), (3.4) allows us to evaluate

$$Z(l) = \frac{V(-l)}{I(-l)} = Z_0 \frac{Z_\mathrm{L} + Z_0 \tanh \gamma l}{Z_0 + Z_\mathrm{L} \tanh \gamma l} \tag{3.8}$$

for a lossless line with a purely imaginary propagation constant γ. l denotes the spatial distance relative to the position of the load, see Fig. 3.1(b).

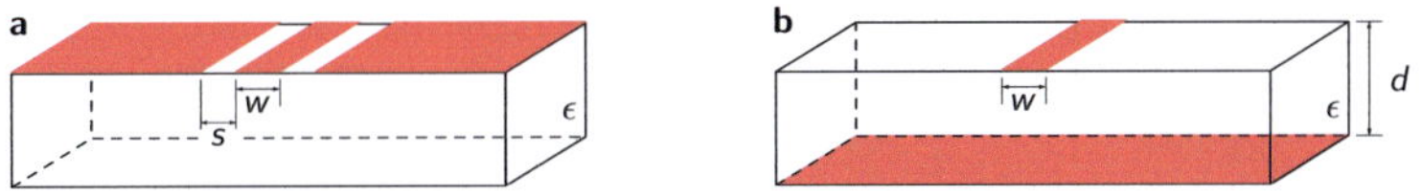

Figure 3.2: Planar implementations of transmission lines (a) Coplanar line with central conductor of width w and ground reference in the device plane at a spacing s. The substrate has a dielectric constant $\epsilon \sim 11$ for intrinsic silicon and sapphire, which are substrate materials commonly used for fabricating superconducting quantum circuits. (b) Microstrip line with single conductor of width w and a ground plane on the backside of the substrate. The thickness d of the substrate gives the distance to the ground reference.

3.2 Circuit implementation of transmission lines

In practice, transmission lines are realized by coaxial lines, where a central conductor at a potential V_0 is surrounded by a concentric shield connected to ground. The distance between inner and outer conductors together with the dielectric constant ϵ of the gap material defines all the quantities sketched in Fig. 3.1(a) and therefore the characteristic impedance Z_0 of the coaxial line.

In order to incorporate transmission lines in planar circuit designs, it is the goal to mimic the properties of a coaxial line as good as possible in a planar geometry. Among others, the coplanar line and the microstrip line are the most popular types of planar transmission lines since they can be easily fabricated by lithography [Poz12].

Figure 3.2(a) shows a schematic of a coplanar transmission line of width w and spacing s to its ground reference on either side. The ratio w/s and the dielectric constant ϵ of the substrate material defines the impedance of the line. A characteristic impedance of $50\,\Omega$ can be achieved for $w/s \sim 5/3$ and $\epsilon_r \sim 11$ [Sim01], as is the case for intrinsic silicon and sapphire. For a substrate thickness that is much larger than the scaled gap size $s\epsilon$, a ground plane on the backside of the substrate has no influence since the fields are mainly confined within the gap region in the device plane. The scaling with ϵ is due to the change of the electric displacement field in a dielectric. A substrate of thickness d and dielectric constant ϵ therefore corresponds to a vacuum equivalent substrate of thickness d/ϵ.

In contrast, a transmission line in microstrip geometry does not have its ground reference in the device plane but at the backside of the substrate at a distance d. The characteristic impedance is dependent the ratio w/d. For a substrate dielectric constant $\epsilon_r \sim 11$, a characteristic impedance of $50\,\Omega$ can be achieved for $w \sim d$ [Poz12].

The microstrip geometry features an increased mode volume due to the larger dimensions between centre conductor and ground plane such that electric fields are smaller which potentially decreases loss from surface and interface defects. Since field are less confined, as compared to the coplanar architecture, radiative loss is favoured. Due to the strong field confinement in coplanar lines, crosstalk between adjacent circuit elements is reduced which benefits the scalability of circuits. Both geometries are employed in this work.

3.3 Microwave network characterization with the scattering matrix

A complete description of a microwave network with an arbitrary number of ports is provided by the scattering matrix $\hat{S}$ [Poz12]. The matrix element S_{ij} relates the wave of voltage V_i^- reflected from port i to the incident wave of voltage V_j^+ travelling into port j,

$$S_{ij} = \left. \frac{V_i^-}{V_j^+} \right|_{V_k^+ = 0 \, \forall k \neq j} . \tag{3.9}$$

The incident waves on all other ports must be set to zero. The scattering matrix will be used in this thesis to characterise single-port and two-port networks. For a single-port network, we obtain the reflection matrix element

$$S_{11} \equiv \Gamma = \frac{V_1^-}{V_1^+} \tag{3.10}$$

as defined in Eq. (3.7). For two-port networks we can write $\hat{S}$ in matrix form

$$\begin{pmatrix} V_1^- \\ V_2^- \end{pmatrix} = \begin{pmatrix} S_{11} & S_{12} \\ S_{21} & S_{22} \end{pmatrix} \begin{pmatrix} V_1^+ \\ V_2^+ \end{pmatrix} . \tag{3.11}$$

If a microwave network does not contain any active devices or ferrites, the network is called reciprocal and the scattering matrix $\hat{S}$ is symmetric such that $S_{11} = S_{22}$ and $S_{12} = S_{21}$ [Poz12].

Experimentally, a vector network analyser (VNA) allows us to characterise any multi-port network. By using two VNA ports, all scattering matrix elements S_{ij} of a network can be measured successively while terminating all ports $k \neq i, l$ in loads matched to the characteristic impedances of the respective lines at ports k.

3.4 Microwave resonators

Microwave resonators are used to implement harmonic oscillator modes in superconducting circuits and therewith form an integral building block in circuit QED. They typically appear as transmission line resonators with a fixed characteristic impedance Z_0 and length l between two boundary conditions defining the transmission line. Continuous reflections at both ends of the resonator gives rise to the formation of standing waves. The resonance frequency ω_0 is dependent on the length l, with $l \propto \lambda$ and λ the wavelength of the resonating microwave. For $\lambda/2$ resonators, which are used in this thesis, the two boundary conditions are identical and the resonance frequency is

$$\omega_0/2\pi = \frac{v_\mathrm{p}}{\lambda} = \frac{mc}{\sqrt{\epsilon_\mathrm{eff}}2l}, \tag{3.12}$$

with the phase velocity $v_\mathrm{p} = c/\sqrt{\epsilon_\mathrm{eff}}$ in the medium of effective dielectric constant ϵ_eff and the speed of light c in vacuum. m counts the mode number of the transmission line resonator with $l = m\lambda/2$, $m \in \mathbb{N}^+$, yielding an infinite number of modes spaced by the resonance frequency ω_0.

The treatment of transmission line resonators is analogous to the one for lumped-element LCR resonators of circuit theory [Poz12]. In the following, we will therefore rather use lumped-element equivalent parameters for resonator characterization instead of referring to the quantities per unit length introduced in Fig. 3.1(a). L denotes the impedance, C the capacitance and R the resistance of an equivalent lumped-element circuit.

3.4.1 Definition of quality factors

Quality factors are a measure for the losses a resonator of resonance frequency ω_0 experiences. A generic quality factor Q is defined as

$$Q = \omega_0 \frac{E_\mathrm{tot}}{P_\mathrm{loss}} = \frac{\omega_0}{\kappa}, \tag{3.13}$$

where E_tot is the total energy stored in the resonator and P_loss is the dissipated power or energy loss rate [Poz12]. According to the common convention, κ denotes the inverse photon lifetime of a linear resonator, defined as the full width at half maximum of a Lorentzian response in frequency space. Via Fourier transformation one can see that this means the cavity relaxes to its groundstate at a rate of $\kappa/2$. In the following we will refer to κ as a decay rate.

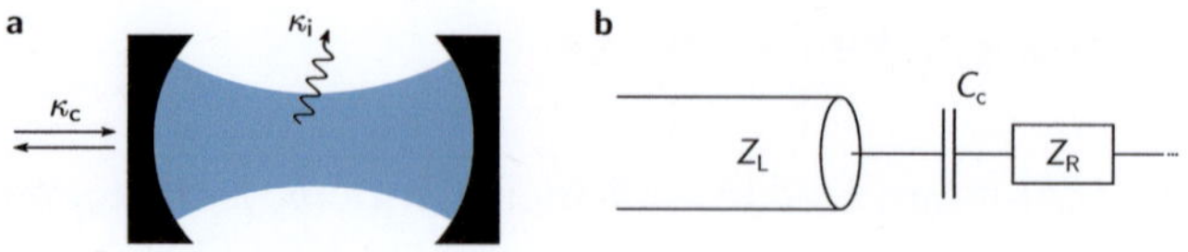

Figure 3.3: Illustration of quality factors and decay rates (a) The coupling decay rate $\kappa_c = \omega_0/Q_c$ and the internal decay rate $\kappa_i = \omega_0/Q_i$ in a resonating cavity yield the total decay rate $\kappa = \kappa_i + \kappa_c$. κ_c is defined separately for each single port of the network and adds to the total decay. (b) Schematic circuit for the derivation of the coupling quality factor Q_c.

Loaded quality factor

The loaded quality factor Q_L accounts for the total loss of a resonating circuit [Poz12]. It is defined as the reciprocal sum of internal quality factor Q_i and coupling quality factor Q_c, as defined below.

$$Q_L^{-1} = Q_i^{-1} + Q_c^{-1} \tag{3.14}$$

Accordingly, the respective decay rates for internal loss κ_i and for external loss κ_c add up to the total loaded decay rate κ, see Fig. 3.3(a).

Table 3.1: Summary of coupling regimes

over-coupled	$Q_c < Q_i$
critical coupling	$Q_c = Q_i$
under-coupled	$Q_c > Q_i$

The relative contributions of dissipation due to internal loss and due to coupling to external parts of the circuit define the coupling regime of the resonating circuit. Table 3.1 summarizes the different coupling regimes. In case $Q_i \gg Q_c$, we find $Q_L \sim Q_c$, which is a common scenario in superconducting quantum circuits.

In the following, we present formulas for resonant circuits formed by a series network of lumped-element components. Formulas for the parallel lumped-element resonant circuit can be obtained in analogy.

Internal quality factor

The internal quality factor Q_i describes intrinsic losses of a resonator for instance via dipole radiation into the vacuum or dielectric and paramagnetic loss at surface and interface defects of metal films. For the series resonant LCR circuit, it is defined as

$$Q_\mathrm{i} = \omega_0 \frac{E_\mathrm{tot}}{P_\mathrm{loss}} = \frac{\omega_0 L}{R} = \frac{1}{\omega_0 RC}. \tag{3.15}$$

P_loss here only accounts for dissipated power via the internal resistor R.

Microwave resonators built with superconducting metals are appealing in particular since they feature very small intrinsic loss. Due to the superconducting gap that opens in the quasi-particle density of states, Cooper pair conduction is dissipation free, rendering the internal quality limited by geometric or surface properties of the resonator. In order to avoid quasi-particle excitations, the superconducting gap must be much larger than the resonance frequency ω_0 of the resonator.

Coupling quality factor

The external quality factor, or coupling quality factor Q_c accounts for the dissipation of a resonator via its coupling to external circuit components. Q_c is therefore closely related to the coupling strength of the resonator to its environment.

In the following we restrict our analysis to the case where resonators are capacitively coupled to external circuitry. Using the definition

$$Q_\mathrm{c} = \omega_0 \frac{E_\mathrm{tot}}{P_\mathrm{diss}}, \tag{3.16}$$

where $E_\mathrm{tot} = \frac{1}{2}\tilde{C}V^2 = \frac{1}{4}\tilde{C}|V|^2$ is the total energy stored in the resonator with maximum voltage $|V|$ and root-mean-square voltage $\frac{1}{\sqrt{2}}|V|$. The total capacitance $\tilde{C} = \pi(\omega Z_\mathrm{R})^{-1}$ of an open-ended $\lambda/2$ resonator can be found from circuit quantization [Pal10], with resonator characteristic impedance Z_R. $P_\mathrm{diss} = VI = \frac{1}{2}|V|^2\Re[Z_\mathrm{ex}^{-1}]$ denotes the dissipated power at a single port of the resonator via the coupling capacitance C_c into a load of impedance Z_L, summarized in an external impedance $Z_\mathrm{ex} = 1/i\omega C_\mathrm{c} + Z_\mathrm{L}$. See Fig. 3.3(b) for a schematic sketch of the considered circuit. With the approximation $Z_\mathrm{L} \ll (\omega C_\mathrm{c})^{-1}$, Eq. (3.16) becomes [Maz04; Sch07]

$$Q_\mathrm{c} = \frac{\pi}{2Z_\mathrm{R}\Re[Z_\mathrm{ex}^{-1}]} = \frac{\pi}{2Z_\mathrm{R}} \frac{Z_\mathrm{L}^2 + (\omega C_\mathrm{c})^{-2}}{Z_\mathrm{L}} \approx \frac{\pi}{2Z_\mathrm{R} Z_\mathrm{L} \omega^2 C_\mathrm{c}^2}. \tag{3.17}$$

Employing the definition of the normalized susceptance [Poz12]

$$b_\mathrm{c} = \sqrt{Z_\mathrm{R} Z_\mathrm{L}}\,\omega C_\mathrm{c} \tag{3.18}$$

of the coupling capacitor, Eq. (3.17) can be rewritten as

$$Q_c = \frac{\pi}{2b_c^2}. \tag{3.19}$$

It is important to point out that the definition of the coupling quality factor Q_c is dependent on the load impedance Z_L that is connected to the resonator. In particular, it is convenient to define a separate coupling quality factor for each external port via which the resonator couples to its environment.

3.4.2 Impedance of a capacitively coupled resonant circuit

Calculating the total impedance of a capacitively coupled resonator to a external load provides a very elegant method to relate scattering matrix elements to the frequency response of many frequently used resonator networks. To this end, we derive the input impedance Z_{in} of an open-ended $\lambda/2$ transmission line resonator of length l, that couples to an external load of impedance Z_L through a capacitor C_c at one of its ends. See Fig. 3.4(a) for a schematic sketch of the circuit. Our goal is to summarize the resonator and its coupling capacitor to an input impedance Z_{in} that is seen by the transmission line. The resulting effective circuit is depicted in Fig. 3.4(b).

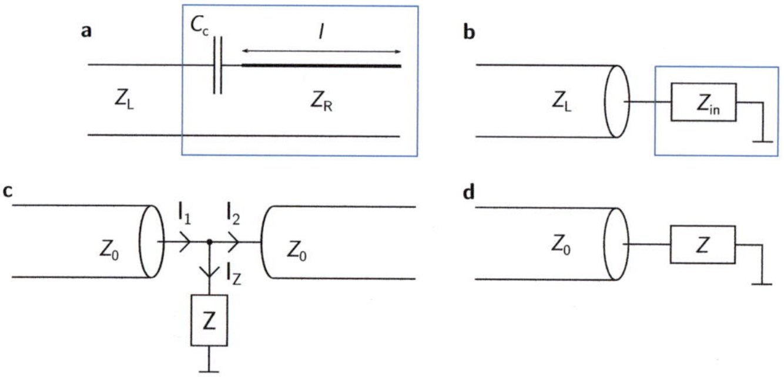

Figure 3.4: Schematics yielding impedance formulas (a) An open-ended $\lambda/2$ resonator with impedance Z_R and of length l that couples to a transmission line of impedance Z_L via a capacitor C_c at one of its ends. Resonator impedance and the coupling are summarized in an effective input impedance Z_{in} as seen by the transmission line, see (b). The blue rectangles in (a), (b) are therewith equivalent. (c) Notch type configuration. An impedance Z couples to a transmission line of characteristic impedance Z_0 as a parallel scatterer. (d) Single-port reflection scenario. An impedance Z couples to an open-ended transmission line of characteristic impedance Z_0.

Following the treatment in Ref. [Poz12], we can write the input impedance Z_{in} using Eq. (3.8) as

$$Z_{\text{in}}(\omega) = \left(\frac{1}{\mathrm{i}\omega C_{\text{c}}} + \frac{Z_{\text{R}}}{\tan\gamma l}\right) \propto -\mathrm{i}\left(\tan\gamma l + Z_{\text{R}}\omega C_{\text{c}}\right). \tag{3.20}$$

Requesting $\Im Z_{\text{in}} = 0$ for the resonance condition, performing a Taylor expansion around resonance $\omega \approx \omega_0$ and inserting $l = \lambda/2$ for a $\lambda/2$ resonator yields

$$Z_{\text{in}}(\omega) = Z_{\text{R}}\left(\frac{\pi}{2Q_{\text{i}}Z_{\text{R}}^2\omega^2C_{\text{c}}^2} + \mathrm{i}\frac{\pi\Delta\omega}{\omega_0 Z_{\text{R}}^2\omega^2C_{\text{c}}^2}\right), \tag{3.21}$$

valid in the vicinity of the resonance frequency ω_0, $\Delta\omega = \omega - \omega_0 \ll \omega_0$. The input impedance is stated relative to the characteristic impedance Z_{R} of the resonator.

The input impedance in Eq. (3.21) takes the form of a series LCR circuit [Poz12] near resonance

$$Z_{\text{ser}} = R\left(1 + 2\mathrm{i}Q_{\text{i}}\frac{\Delta\omega}{\omega_0}\right) \tag{3.22}$$

noting the relations $Q_{\text{c}} = \omega_0 L/Z_{\text{L}}$ and $Q_{\text{i}} = \omega_0 L/R$.

3.4.3 Scattering matrix elements for relevant resonator networks

In this section we provide relations between scattering matrix elements and response functions of microwave resonators in one- and two-port networks. These allow us to fit experimental data and to extract characteristic resonator properties. We focus the presented derivations on the notch type configuration and single-port reflection from a resonator, since these scenarios are used in this thesis. A complete summary of the obtained results including the in-line configuration is given in Table 3.2 at the end of this section. We base our treatment on the input impedance of the investigated resonating circuit rather than applying input-output theory [Pal10], which is an equivalent approach. Our results are in agreement with literature [Göp+08; Gao08; Kha+12; Kra16; Pal10; PA98; Poz12; Pro+15; Sha+11].

Notch type configuration

In order to calculate reflection S_{11}^{n} and transmission S_{21}^{n} scattering matrix elements for a notch type configuration, we consider the circuit depicted in Fig. 3.4(c). In this scenario, an impedance Z couples to a continuous transmission line of characteristic impedance Z_0 as a parallel scatterer. The impedance Z sees a load $Z_{\text{L}} = Z_0/2$ due to the two branches of the transmission line such that we can rewrite Eq. (3.21) in terms of $Q_\mathrm{c} = \pi / Z_\mathrm{R} Z_0 \omega_0^2 C_\mathrm{c}^2$ as

$$Z_\mathrm{in}(\omega) = Z_0 \left(\frac{Q_\mathrm{c}}{2Q_\mathrm{i}} + \mathrm{i}\frac{\Delta\omega}{\omega_0} Q_\mathrm{c} \right). \tag{3.23}$$

In order to find relations between the scattering matrix elements and the input impedance Z_in we make use of the table provided in Ref. [Poz12], p. 190, 192 on conversions between the scattering matrix elements and the transmission (ABCD) matrix elements for two-port networks. Exemplarily, we derive the relation between the transmission scattering matrix element S_{21}^n to the input impedance of a notch type resonator [Lep+17]. We denote V_i^+ (V_i^-) the voltage contribution of a wave propagating from left to right (right to left) on site $i = 1, 2$, where 1 is left of the scatterer and 2 is right of the scatterer, according to Fig. 3.4(c). Net currents I_i are labelled accordingly and I_Z denotes the current flowing in the branch of the scatterer. From Kirchhoff's rules, we obtain the relations

$$V_1^+ + V_1^- = V_2^+ + V_2^- \tag{3.24}$$

$$I_1 = I_2 + I_Z \tag{3.25}$$

$$\Leftrightarrow \frac{1}{Z_0}\left(V_1^+ - V_1^-\right) = \frac{1}{Z_0}\left(V_2^+ - V_2^-\right) + \frac{1}{Z}\left(V_1^+ - V_1^-\right) \tag{3.26}$$

and therefore the scattering parameter

$$S_{21}^\mathrm{n} \equiv \frac{V_2^+}{V_1^+} = \frac{1}{1 + \frac{Z_0}{2Z}} = \frac{2Z}{2Z + Z_0} \tag{3.27}$$

where Z_0 is the characteristic impedance of the transmission line. This result is in agreement with Ref. [Poz12]. In analogy, we find

$$S_{11}^\mathrm{n} = \frac{Z_0}{Z_0 + 2Z}, \tag{3.28}$$

such that we obtain the relation $S_{11}^\mathrm{n} + S_{21}^\mathrm{n} = 1$. Note that this is a result specific for the notch configuration.

Inserting Eq. (3.23) into Eq. (3.27) yields the complex transmission scattering parameter [Gao08; Kha+12; Pro+15]

$$S_{21}^\mathrm{n}(\omega) = 1 - \frac{Q_\mathrm{L}/Q_\mathrm{c}}{1 + 2iQ_\mathrm{L}\frac{\Delta\omega}{\omega_0}}. \tag{3.29}$$

The absolute square

$$|S_{21}^\mathrm{n}|^2(\omega) = \left(\Re[S_{21}^\mathrm{n}]\right)^2 + \left(\Im[S_{21}^\mathrm{n}]\right)^2 = 1 - \frac{1 - (Q_\mathrm{L} - Q_\mathrm{c})^2/Q_\mathrm{c}^2}{1 + 4Q_\mathrm{L}^2\left(\frac{\Delta\omega}{\omega_0}\right)^2} \tag{3.30}$$

takes the form of an inverse Lorentzian distribution of full width at half maximum $\kappa = \omega_0/Q_L$, reflecting the dip structure in the transmitted magnitude response. The phase of the complex scattering parameter in Eq. (3.29) becomes

$$\arg S^n_{21}(\omega) = \arctan\left(\frac{\Im[S^n_{21}]}{\Re[S^n_{21}]}\right) = \arctan\left(\frac{2Q_L\Delta\omega/\omega_0}{\frac{Q_c}{Q_L} - 1 + 4Q_LQ_c\left(\frac{\Delta\omega}{\omega_0}\right)^2}\right) \tag{3.31}$$

$$= -\arctan\left(2Q_L\frac{\Delta\omega}{\omega_0}\right) - \arctan\left(2\frac{Q_LQ_c}{Q_L - Q_c}\frac{\Delta\omega}{\omega_0}\right). \tag{3.32}$$

We used *arctan*'s addition theorem

$$\arctan(u) + \arctan(v) = \arctan\left(\frac{u+v}{1-uv}\right), \tag{3.33}$$

valid for $uv < 1$. $4Q_L^2Q_c(Q_L - Q_c)^{-1}(\Delta\omega/\omega_0)^2 < 1$ is fulfilled since $\Delta\omega/\omega_0 \ll 1$. Phase data can be used to fit purely Q_L for inverting S^n_{21} and shifting the coordinate system by 1 on the real axis, $S^n_{21} \to 1 - S^n_{21}$, as pointed out in [Pro+15].

$$\arg(1 - S^n_{21}) = \arctan\left(\frac{\Im[1 - S^n_{21}]}{\Re[1 - S^n_{21}]}\right) = -\arctan\left(2Q_L\frac{\Delta\omega}{\omega_0}\right), \tag{3.34}$$

and its derivative with respect to ω becomes

$$\frac{d}{d\omega}\arg(1 - S_{21})(\omega) = \frac{-2Q_L/\omega_0}{1 + 4Q_L^2\left(\frac{\Delta\omega}{\omega_0}\right)^2} = \frac{\frac{\kappa}{2}}{\left(\frac{\kappa}{2}\right)^2 + \Delta\omega^2}, \tag{3.35}$$

which recovers an analog Lorentzian distribution as derived in Eq. (3.30). As we point out in Sec. 3.4.3, this provides a robust method to extract resonator characteristics.

From Eq. (3.28) and using Eq. 3.23 we can immediately write down the reflection matrix element

$$S^n_{11} = \frac{Q_L/Q_c}{1 + 2iQ_L\frac{\Delta\omega}{\omega_0}} \tag{3.36}$$

and therewith

$$|S^n_{11}| = \frac{(Q_L/Q_c)^2}{1 + 4Q_L^2\left(\frac{\Delta\omega}{\omega_0}\right)^2}, \quad \arg S^n_{11} = -\arctan\left(2Q_L\frac{\Delta\omega}{\omega_0}\right). \tag{3.37}$$

Single-port reflection

In the single-port scenario, we have $Z_L = Z_0$ according to the circuit in Fig. 3.4(d) and can identify $Q_c = \pi/2Z_RZ_0\omega_0^2C_c^2$. The input impedance becomes

$$Z_{in}(\omega) = Z_0\left(\frac{Q_c}{Q_i} + i\frac{\Delta\omega}{\omega_0}2Q_c\right). \tag{3.38}$$

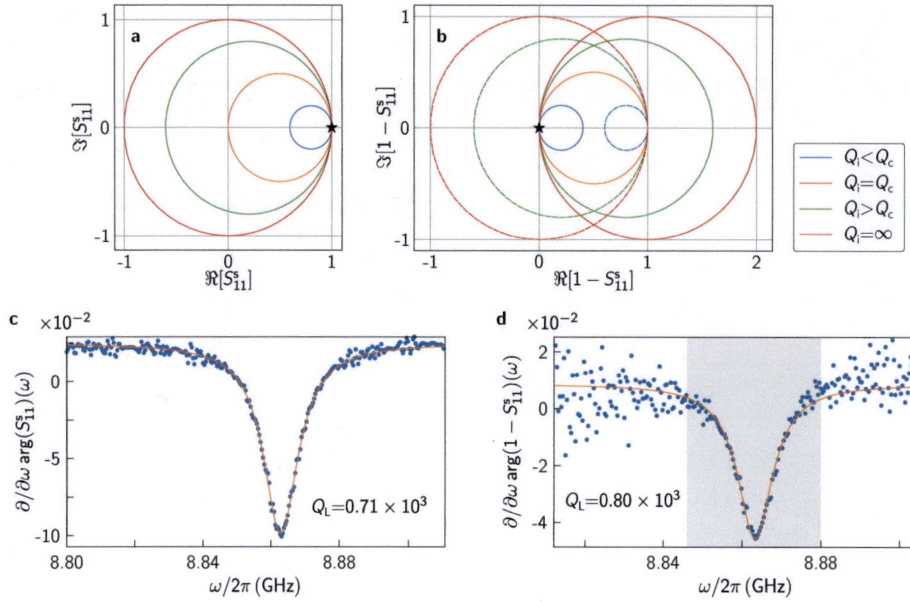

Figure 3.5: Single port reflection from a resonator. (a) Plot of the scattering parameter S^s_{11} in the complex plane. The off-resonance point is marked with a black star and different colours correspond to various values of Q_i relative to a fixed Q_c. (b) Shifted and inverted complex data $1 - S^s_{11}$ (solid circles) with the original circles (dashed) as shown in (b). The off-resonance point (black star) is shifted to the origin of the complex plane and circles are in the right half plane such that $\arg S^s_{11}$ covers a range of only π irrespective of the coupling regime. (c) Lorentzian fit of the phase data $\partial/\partial\omega \arg S^s_{11}(\omega)$ differentiated with respect to frequency ω, yielding a $Q_L = 0.71 \times 10^3$. (d) Similar Lorentzian fit of the argument of the normalized and shifted reflection data $\partial/\partial\omega \arg(1 - S^s_{11})$, yielding a $Q_L = 0.80 \times 10^3$. For the fit in in (e), only points in the grey area are considered.

The single-port reflection coefficient S^s_{11} for looking into a transmission line of characteristic impedance Z_0, terminated in a load $Z = Z_{in}$ can be found to be

$$S^s_{11} = \frac{V_0^-}{V_0^+} = \frac{Z - Z_0}{Z + Z_0} \tag{3.39}$$

such that

$$S^s_{11}(\omega) = 1 - \frac{2Q_L/Q_c}{1 + 2iQ_L\frac{\Delta\omega}{\omega_0}}. \tag{3.40}$$

Correspondingly, find

$$|S_{11}^{s}|^2 = 1 - \frac{1 - \frac{(2Q_L - Q_c)^2}{Q_c^2}}{1 + 4Q_L^2 \left(\frac{\Delta\omega}{\omega_0}\right)^2}, \tag{3.41}$$

$$\arg S_{11}^{s} = -\arctan\left(2Q_L \frac{\Delta\omega}{\omega_0}\right) - \arctan\left(2 \frac{Q_L Q_c}{2Q_L - Q_c} \frac{\Delta\omega}{\omega_0}\right). \tag{3.42}$$

For a coupling limited resonator with $Q_L = Q_c$ and $Q_i \to \infty$ we find $|S_{11}^{s}|^2(\omega) = 1$, which is intuitively clear since all the energy put into the resonator will come back eventually in the absence of internal loss. In the complex plane, this corresponds to a circle that is centred around the origin, see Fig. 3.5(a). In the same limit, $Q_i \to \infty$, Eq. (3.42) reduces to

$$\arg S_{11}^{s} = -2\arctan\left(2Q_L \frac{\Delta\omega}{\omega_0}\right), \tag{3.43}$$

accounting for a 2π phase roll-off. The respective circle in the complex plane encloses the origin of the coordinate system. These results are in agreement with Ref. [Pal10]. Note that we can recover a similar formula as Eq. (3.43) for all coupling regimes when evaluating

$$\arg(1 - S_{11}^{s}) = \arctan\left(\frac{\Im[1 - S_{11}^{s}]}{\Re[1 - S_{11}^{s}]}\right) = -\arctan\left(2Q_L \frac{\Delta\omega}{\omega_0}\right). \tag{3.44}$$

The shift along the real axis can be motivated as follows: Since the co-domain of the *arctan* function is only π, the scattering data needs to be shifted into the positive complex half-plane such that the phase range of scattering data is limited to π for all coupling regimes. Since the magnitude is normalized to unity, a shift by unity is required. The additional factor of 2 in Eq. (3.43) reflects the 2π co-domain of S_{11}^{s}, as the origin lies within the circle in the over-coupled regime. This is illustrated in Fig. 3.5(b).

Given that the magnitude signal $|S_{11}^{s}|^2$ is very small for strongly over-coupled resonators, a robust method to extract resonator characteristics is provided by fitting the frequency derivative of the phase to a Lorentzian distribution. In Fig. (3.5)(c), (d) we compare Lorentzian fits to the derivative of the phase quadrature $\partial/\partial\omega \arg S_{11}^{s}$ and to the inverted and shifted phase quadrature, $\partial/\partial\omega \arg(1 - S_{11}^{s})$. While the latter is the analytically exact procedure, it requires an elaborate post-processing of measured data, such as a normalization and a slope subtraction in the phase signal, which is physically hard to justify. Comparing the extracted loaded quality factors demonstrates the reliability of the straightforward method. The internal quality factor in the example is $\sim 20 \times 10^3$.

Table 3.2: Summary of relations between scattering matrix elements and frequency response of relevant one- and two-port resonator networks In the in-line configuration, a resonator is defined by introducing two gaps in a continuous transmission line. The reflection case is analog to single-port reflection, with individual coupling quality factors $Q_{\mathrm{c,i}}$ for port $i = 1, 2$, respectively. For equal coupling capacitors at either port, we can recover $S_{11}^{\mathrm{il,i}} = S_{11}^{\mathrm{s}}$ for $Q_{\mathrm{c}} = Q_{\mathrm{c,i}}/2$. The in-line transmission matrix element S_{21}^{il} is provided with respect to a the total coupling quality factor $Q_{\mathrm{c},\Sigma}^{-1} = Q_{\mathrm{c},1}^{-1} + Q_{\mathrm{c},2}^{-1}$.

	scattering matrix elem.	magnitude squared	phase
notch	$S_{21}^{\mathrm{n}} = 1 - \frac{Q_{\mathrm{L}}/Q_{\mathrm{c}}}{1+2iQ_{\mathrm{L}}\frac{\Delta\omega}{\omega_0}}$	$\lvert S_{21}^{\mathrm{n}}\rvert^2 = 1 - \frac{1-(Q_{\mathrm{L}}-Q_{\mathrm{c}})^2/Q_{\mathrm{c}}^2}{1+4Q_{\mathrm{L}}^2\left(\frac{\Delta\omega}{\omega_0}\right)^2}$	$\arg S_{21}^{\mathrm{n}} = -\arctan\left(2Q_{\mathrm{L}}\frac{\Delta\omega}{\omega_0}\right) - \arctan\left(2\frac{Q_{\mathrm{L}}Q_{\mathrm{c}}}{Q_{\mathrm{L}}-Q_{\mathrm{c}}}\frac{\Delta\omega}{\omega_0}\right)$
	$S_{11}^{\mathrm{n}} = \frac{Q_{\mathrm{L}}/Q_{\mathrm{c}}}{1+2iQ_{\mathrm{L}}\frac{\Delta\omega}{\omega_0}}$	$\lvert S_{11}^{\mathrm{n}}\rvert^2 = \frac{(Q_{\mathrm{L}}/Q_{\mathrm{c}})^2}{1+4Q_{\mathrm{L}}^2\left(\frac{\Delta\omega}{\omega_0}\right)^2}$	$\arg S_{11}^{\mathrm{n}} = -\arctan\left(2Q_{\mathrm{L}}\frac{\Delta\omega}{\omega_0}\right)$
single port	$S_{11}^{\mathrm{s}} = 1 - \frac{2Q_{\mathrm{L}}/Q_{\mathrm{c}}}{1+2iQ_{\mathrm{L}}\frac{\Delta\omega}{\omega_0}}$	$\lvert S_{11}^{\mathrm{s}}\rvert^2 = 1 - \frac{1-(2Q_{\mathrm{L}}-Q_{\mathrm{c}})^2/Q_{\mathrm{c}}^2}{1+4Q_{\mathrm{L}}^2\left(\frac{\Delta\omega}{\omega_0}\right)^2}$	$\arg S_{11}^{\mathrm{s}} = -\arctan\left(2Q_{\mathrm{L}}\frac{\Delta\omega}{\omega_0}\right) - \arctan\left(2\frac{Q_{\mathrm{L}}Q_{\mathrm{c}}}{2Q_{\mathrm{L}}-Q_{\mathrm{c}}}\frac{\Delta\omega}{\omega_0}\right)$
in-line	$S_{21}^{\mathrm{il}} = \frac{Q_{\mathrm{L}}/Q_{\mathrm{c},\Sigma}}{1+2iQ_{\mathrm{L}}\frac{\Delta\omega}{\omega_0}}$	$\lvert S_{21}^{\mathrm{il}}\rvert^2 = \frac{(Q_{\mathrm{L}}/Q_{\mathrm{c},\Sigma})^2}{1+4Q_{\mathrm{L}}^2\left(\frac{\Delta\omega}{\omega_0}\right)^2}$	$\arg S_{21}^{\mathrm{il}} = -\arctan\left(2Q_{\mathrm{L}}\frac{\Delta\omega}{\omega_0}\right)$
	$S_{11}^{\mathrm{il,i}} = 1 - \frac{2Q_{\mathrm{L}}/Q_{\mathrm{c,i}}}{1+2iQ_{\mathrm{L}}\frac{\Delta\omega}{\omega_0}}$	$\lvert S_{11}^{\mathrm{il,i}}\rvert^2 = 1 - \frac{1-(2Q_{\mathrm{L}}-Q_{\mathrm{c,i}})^2/Q_{\mathrm{c,i}}^2}{1+4Q_{\mathrm{L}}^2\left(\frac{\Delta\omega}{\omega_0}\right)^2}$	$\arg S_{11}^{\mathrm{il,i}} = -\arctan\left(2Q_{\mathrm{L}}\frac{\Delta\omega}{\omega_0}\right) - \arctan\left(2\frac{Q_{\mathrm{L}}Q_{\mathrm{c,i}}}{2Q_{\mathrm{L}}-Q_{\mathrm{c,i}}}\frac{\Delta\omega}{\omega_0}\right)$

4 Experimental methods

This chapter presents the experimental methods used in the experiments carried out during this thesis. We provide the basic concepts employed in sample fabrication with a focus on patterning Josephson tunnel junctions. A detailed list of parameters used during fabrication can be found in the Appendix. In order to reveal quantum effects with the fabricated samples, they are cooled to millikelvin temperatures, where classical noise that destroys quantum coherence is strongly suppressed. We describe the utilized cryogenic setup and discuss the working principle of a dilution refrigerator. Qubit measurements are carried out in the microwave regime either by spectroscopy or by applying time-resolved pulses. The basic framework of the microwave setup and equipment used is described in the following. Specific techniques and adaptations in different experiments are discussed in the subsequent result chapters.

4.1 Sample fabrication

Experiments in this thesis were carried out with samples that were in-house fabricated at the Nanostructure Service Laboratory (NSL) of the Karlsruhe Institute of Technology.

We make use of intrinsic silicon substrates of 380 µm thickness and dielectric constant $\epsilon_r = 11.5$ for all qubit samples and use sapphire substrates at a thickness of 440 µm and $\epsilon_r = 10$. The substrate is rinsed in piranha solution to remove organic residues and the surface is plasma cleaned prior to metal deposition in a 'descum' process using a mixture of argon and oxygen.

4.1.1 Josephson junction fabrication

Josephson tunnel junctions with an overlap area of about 100 nm × 220 nm were fabricated by making use of electron beam lithography and shadow angle evaporation. In order to achieve a clean oxide interlayer between two aluminium (Al) electrodes, it is advantageous to performed the deposition in-situ, without breaking the vacuum for the secondary lithography step. This is facilitated by the shadow angle evaporation technique. Here, an overhanging bridge is formed by making use of a double resist stack and the two electrodes are deposited at two different angles relative to the sample surface. We make use of the Dolan bridge technique [Dol77], where a free standing resist bridge is created by a stack of LOR/PMMA resists. After evaporation of the first metal layer, a controlled amount of oxygen is allowed into the chamber for a fixed amount of time and given partial pressure, where the pressure is adjusted by a continuous oxygen flow rate and dynamical pumping. The oxygen exposure leads to the formation of a $\sim$ 1 nm thin layer of AlO_x, later on used as the tunnel barrier of the Josephson junction. We use electrode film thicknesses of 30 nm and 50 nm, respectively, resulting in a total Al film thickness of 80 nm across the entire chip. In a final lift-off process, the resist stack is removed.

4.1.2 Film deposition

Film deposition is primarily performed by evaporation of Al from a thermal source[1] at a background chamber pressure of about 3×10^{-8} mbar.

4.1.3 Optical lithography

For optical lithography, we use the photo resist AZ5214E. It is applied with a spin coater by first covering the complete chip homogeneously at a low spinning frequency and subsequently adjusting its thickness to $\sim$ 1 µm with a higher spinning frequency. Mechanical stability is provided by a successive baking step.

Optical resist structuring is performed by using a chrome mask that is previously written with direct laser lithography. The photo mask is aligned on top of the resist layer with a Carl Süss mask aligner and exposed at a wavelength of 365 nm in hard contact mode. Resist development is carried out by immersing into the TMAH based solvent MIF726, which dissolves exposed regions of resist.

Throughout this thesis, we made use of negative lithography in combination with a lift-off process. In this process, resist is structured on a clean substrate prior to covering the complete chip with a metal film. Thereby, metal is deposited on regions of the substrate that are not covered with resist. In the subsequent lift-off process, resist residues are stripped such that unwanted metallizations are removed.

[1] Plassys Bestec, model MEB550S

Figure 4.1: Cryogenic setup and sample box (a) Aluminium sample box and lid as used in this thesis. (b) View of the base stage of the cryostat, reaching a temperature of typically 30 mK. Samples are enclosed in a cryoperm shield for magnetic shielding (dark box). To the left, the cryogenic circulators are visible. (c) Wet dilution refrigerator without thermal radiation shields. As indicated, the temperature of the stages decreases from top to bottom.

4.2 Cryogenic setup

Samples hosting the fabricated quantum circuits are fixed inside an Al sample box with silver glue and electronically connected by Al wire bonds to a 50 Ω matched microstrip transmission line on a printed circuit board element. The transmission line is then soldered to SMA plugs that connect to the coaxial wiring in the cryostat. See Fig. 4.1(a) for a photograph of the sample box. The Al sample box is thermally anchored to the base plate of a cryostat reaching a temperature of $\sim 30\,\mathrm{mK}$ and enclosed in a cryoperm case for additional magnetic shielding, see Fig. 4.1(b).

We operated a wet dilution refrigerator manufactured partially by High Precision Devices (HPD) and in parts by the in-house workshop, see Fig. 4.1(c).

Liquid ^{4}He at a temperature of 4.2 K can be cooled to about 1.5 K by continuous pumping at its surface and hence reducing its boiling temperature. For this purpose the cryostat contains a reservoir of liquid ^{4}He.

In order to cool below 1.5 K, we make use of a standard technique employing a helium isotope mixture of ^{4}He/^{3}He. The fraction of $\sim 25\%$ of ^{3}He is continuously circulated in a closed cycle by means of a room temperature pump. Warm ^{3}He gas is pre-cooled in a condenser which is thermally connected to a temperature stage of $\sim 1.5\,\mathrm{K}$. The liquefied ^{3}He is passed through a set of high-impedance capillaries and further cooled by ^{3}He coming from the base stage via heat exchangers. It enters the mixing chamber that is thermally anchored to the base plate of the cryostat. Below a critical temperature, the helium isotope mixture in the mixing chamber undergoes a spontaneous phase separation into a ^{3}He rich phase and a dilute phase of $\sim 7\%$ ^{3}He admixture [CL04]. Due to the negative pressure created by the pump at room temperature, ^{3}He from the rich phase is evaporated into the dilute phase such that the cooling power at the phase boundary originates from the latent heat. The ^{3}He leaves the mixing chamber at the dilute phase and enters the still, where it is thermally evaporated and thereby separated from the ^{4}He.

4.3 Microwave setup

Microwave tones and pulses are fed to the cryostat at room temperature $T_{\mathrm{RT}} = 300\,\mathrm{K}$ and are strongly attenuated at stages with temperatures $T_{4\mathrm{K}} = 4\,\mathrm{K}$, $T_{\mathrm{im}} = 0.1\,\mathrm{K}$ and $T_{\mathrm{base}} = 30\,\mathrm{mK}$. Since the resistors in the attenuators are thermally anchored, the electronic noise temperature T_{n} at the base plate is decreased according to

$$T_{\mathrm{n}} = T_{\mathrm{base}} + \frac{T_{\mathrm{im}}}{k_1} + \frac{T_{4\mathrm{K}}}{k_1 k_2} + \frac{T_{\mathrm{RT}}}{k_1 k_2 k_3} = 31.7\,\mathrm{mK} = 1.057 \cdot T_{\mathrm{base}} \tag{4.1}$$

with attenuations $k_1 = k_2 = k_3 = 100$ corresponding to a 20 dB reduction of Johnson-Nyquist noise power, see the lower part in Fig. 4.2. The resulting noise temperature is therefore only about 6% above the base plate temperature such that the signal and in particular thermal noise from the microwave sources and signal generators at room temperature is strongly damped.

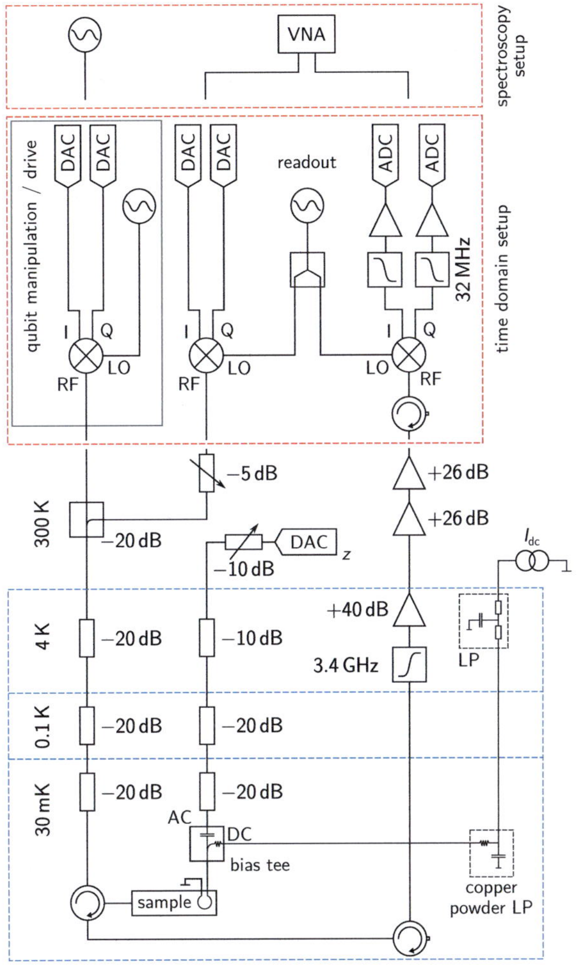

Figure 4.2: Schematic microwave setup used for qubit measurements Dependent on the type of measurement we either connect a spectroscopy setup or a time domain setup (red boxes) to the cryostat (blue box). Circuit elements in between the boxes sit at room temperature and are used in both assemblies. The qubit manipulation and drive part inside the grey box is aptly upgraded for the experiment described in Ch. 6.

In most experiments in this thesis we measure the microwave reflection signal at a single-ended 50 Ω matched transmission line that capacitively couples to further circuitry of the investigated sample. We achieve a splitting of in- and out-going signals by making use of a cryogenic microwave circulator in front of the sample, such that the measurement appears to be transmission type from outside the cryostat.

The measurement signal coming from the sample passes several isolation and filtering stages and is amplified at the 4 K stage by a high electron mobility transistor (HEMT) amplifier[2] with an added noise equivalent temperature of 5.8 K. The HEMT amplifier band is 4 GHz to 16 GHz while the experimentally usable band from 8 GHz to 12 GHz is limited by the cryogenic circulators.

The qubit transition frequency is adjusted by a dc current applied to the on-chip flux coil. High frequency noise is filtered at the 4 K stage with RCR type π-filters at about 25 kHz and on the base plate via a copper powder low pass filter [LU08].

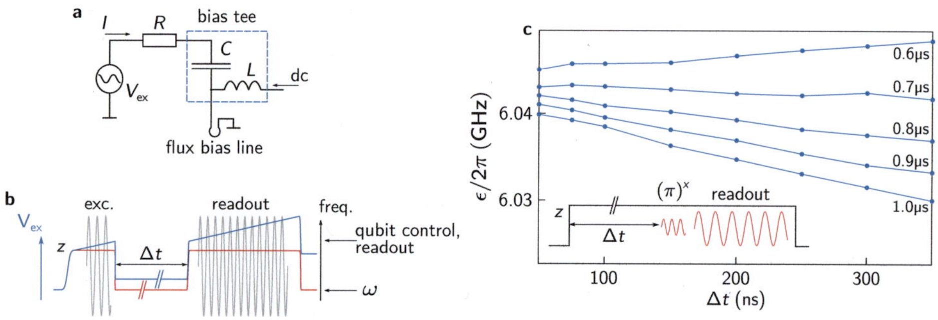

Figure 4.3: Calibration of the bias tee (a) Simplified schematic circuit diagram motivating the required compensation strategy for the finite time constant $\tau = (RC)^{-1}$ in the bias tee. Microwave pulses of amplitude V_{ex} pass a resistor R and charge the capacitor of capacitance C which is part of the bias tee. (b) We apply voltage pulses of amplitude V_{ex} following the blue line to the input of the bias tee such that the resulting current admitted to the on-chip flux bias line follows the desired pulse sequence depicted in red. (c) Typical calibration measurement. The qubit frequency ϵ is measured after increasing detuning times Δt before applying the excitation pulse. We find a constant qubit frequency for choosing $\tau = 0.7\,\mu s$ using a bias tee operated at base temperature.

[2] Low Noise Factory LNF-LNC4_16a sn013b

4.3.1 Bias tee calibration for fast qubit pulsing

Current pulses for fast $\hat{Z}$ pulsing of the qubit are sent through a separate microwave line and combined with the dc bias component by means of a bias tee located at the base plate. It conserves the $50\,\Omega$ impedance matching of the microwave line but contains a decoupling capacitor that is only transparent for the ac component in the applied signal, see the schematic circuit diagram in Fig. 4.3(a). The discharge at a finite time constant $\tau = (RC)^{-1}$ requires a continuous compensation of the applied voltage in order to maintain a constant current that is applied to the flux bias line.

V_{ex} denotes the amplitude of the voltage pulse, R is the line resistance in front of the bias tee and C is the capacitance of the bias tee. From Kirchhoff's law we can write

$$RI + \frac{Q}{C} = V_{\text{ex}}, \tag{4.2}$$

with I the constant current that is admitted to the flux bias line. The condition of constant current follows from requesting

$$\frac{\mathrm{d}I}{\mathrm{d}t} = \frac{1}{R}\frac{\mathrm{d}}{\mathrm{d}t}V_{\text{ex}} - \frac{1}{RC}I \stackrel{!}{=} 0, \tag{4.3}$$

which yields

$$\frac{\mathrm{d}}{\mathrm{d}t}V_{\text{ex}} = \frac{I}{C}. \tag{4.4}$$

Integrating Eq. (4.4) yields

$$V_{\text{ex}} = \frac{1}{C}It + \text{const.} \propto t. \tag{4.5}$$

This shows that the required correction of the externally applied voltage V_{ex} is linear in time t with a slope proportional to the time constant $1/\tau$. See the pulse sequence applied (blue) and the resulting pattern (red) in Fig. 4.3(b). During pulse-off time where $I = 0$, V_{ex} has to be kept constant and no further correction is required.

The described compensation cannot be performed for longer than approximately $1\,\mu\text{s}$ within one continuous pulse sequence, since the output stage of the pulse generator eventually saturates or high currents heat the cryostat. We therefore typically utilize a scheme where fast $\hat{Z}$ pulsing is only required during excitation and readout of a measurement sequence, which scales nicely for perspectively longer recording times Δt.

We calibrated the time constant of the bias tee[3] used in the experiment to be $\tau = 0.7\,\mu\text{s}$.

[3] Tektronix PSPL5547

4.3.2 Spectroscopy

The spectroscopy setup consists of a vector network analyser (VNA)[4] that measures the complex transmission scattering matrix element S_{21}. A continuous microwave tone is admitted to the sample and the dispersive resonator shift can be extracted from the respective response signal. By means of a continuous drive from a separate microwave source, specific resonant transitions in a qubit circuit can be excited. In spectroscopy, the qubit assumes an equal superposition state between its ground state and the driven state which induces a specific dispersive shift on the readout resonator. The mentioned components of the spectroscopy setup are schematically depicted in the top part of Fig. 4.2.

4.3.3 Time resolved measurements

In order to perform time resolved measurements, a phase sensitive setup that generates and records microwave pulses is required. Figure 4.2 schematically shows the basic structure of the time domain setup used in the experiments of this thesis.

Readout of the qubit state is performed by recording the dispersive shift of a readout resonator that couples to the qubit. We apply a strong projective measurement of the qubit $\hat{\sigma}_z$ operator with a readout pulse of 400 ns duration, leading to a circulating photon number in the resonator in excess of the single photon regime. Readout pulses are generated via heterodyne single sideband mixing of a local oscillator (LO) carrier frequency by an IQ frequency of 30 MHz. The shift in resonance frequency of the readout resonator is extracted by acquiring the reflected microwave pulse after down-conversion with the same LO frequency in a second IQ mixer. A complex value for the measured scattering parameter can be reconstructed from the measured signals in I and Q acquired at the IQ frequency after Fourier transformation. This heterodyne frequency conversion scheme eliminates parasitic population of the readout resonator during pulse-off time which would cause photon induced qubit decay.

Data acquisition is performed by recording full time traces of the readout pulses reflected from the sample, which are $\sim 2 \times 10^3$ fold pre-averaged per trace on an analog-to-digital conversion (ADC) acquisition card. Successively, the data is sent to the measurement computer where we extract the IQ quadratures by Fourier transformation. In measurements of the qubit lifetime, its dephasing rate or its time evolution, we typically average over ~ 30 complete traces to obtain a reasonable

[4] Anritsu MS4642B

signal to noise ratio. Due to the single-port reflection scheme, most information is stored in the phase quadrature of the recorded signal since $Q_\mathrm{i} \gg Q_\mathrm{c}$ for the readout resonator, see Sec. 3. Error bars are calculated from pre-averaged data points and represent the standard deviation of the mean.

Qubit manipulation pulses are equally generated by heterodyne single sideband mixing and combined in a directional coupler to be applied to the same transmission line used for readout. See Fig. 4.2 for a detailed schematics of the microwave setup.

We use Anritsu MG37022A and Agilent E8257D signal generators as well as Tektronix AWG7062B and Tabor WX1284C arbitrary waveform generators (DAC). The IQ mixers are Marki IQ0618LXP for the readout and IQ4509LXP, IQ0307LXP for qubit manipulation. All devices in the setup are phase-locked to a common 10 MHz clock reference and triggers are generated by an Agilent 81130A pulse generator.

4.4 Measurement software

The measurement software utilized in this thesis is written in Python 2.7 and is based on the QTLab measurement software mainly written by R. Heeres [Hee]. In 2015, we merged and extended existing measurement scripts at KIT and collected them in a software framework called qkit [qkit15], which was later uploaded to a public repository of the GitHub development platform. Qkit now provides a measurement suite for quantum microwave measurements that is typically controlled via an IPython notebook interface. It features data acquisition and analysis including robust fitting tools, intelligent data storage and visualization as well as server based data monitoring and logging.

5 The concentric transmon qubit

We present a planar qubit design based on a superconducting circuit that we call concentric transmon. While employing a straightforward fabrication process using Al evaporation and lift-off lithography, we reproducibly observe qubit lifetimes and coherence times up to 15 µs. *We systematically characterize loss channels such as incoherent dielectric loss, Purcell decay and radiative losses. The implementation of a gradiometric SQUID loop allows for a fast tuning of the qubit transition frequency and therefore for full tomographic control of the quantum circuit. Due to the large loop size, the presented qubit architecture features a strongly increased magnetic dipole moment as compared to conventional transmon designs. This renders the concentric transmon a promising candidate to establish a site-selective direct* $\hat{Z}$ *coupling between neighbouring qubits, being a pending quest in the field of quantum simulation.* [Bra+16]

5.1 Motivation

Over the last decade there has been a two order of magnitude increase in coherence times of superconducting qubits, which allowed for several major achievements in the pursuit of scalable quantum computation, such as the control and entanglement of multiple qubits [Ste+06; Bar+14]. Further increases in coherence times will eventually allow for building a fault tolerant quantum computer with a reasonable overhead in terms of error correction, as well as implementing novel quantum simulation schemes by accessing wider experimental parameter ranges [Par14]. Apart from the demand for advancing qubit lifetimes and coherence times, other properties become more and more important, such as the scalability of quantum circuits and the coupling versatility between qubits.

The motivation for building the concentric transmon qubit is to add a building block to the quantum toolbox that features sufficient coherence properties for many experiments in the emerging field of quantum simulations while making use of a straightforward and robust fabrication technique. It is the goal to achieve a compact planar device with a confined field distribution that suppresses unwanted crosstalk to spurious modes. This addresses the issue of scalability which was reported as a weakness of superconducting circuits in general [GAN14]. Additionally, we aim for a device featuring a longitudinal magnetic coupling.

While superconducting qubits embedded in a 3D cavity [Pai+11] have shown coherence times in excess of 100 µs [Rig+12] and even 1 ms [Pop+14], this approach imposes some constraints on the tunability and the mutual coupling of individual circuit elements. Since the Josephson junction itself does not limit qubit coherence [Pai+11], comparably long lifetimes can also be achieved in a planar geometry by careful circuit engineering.

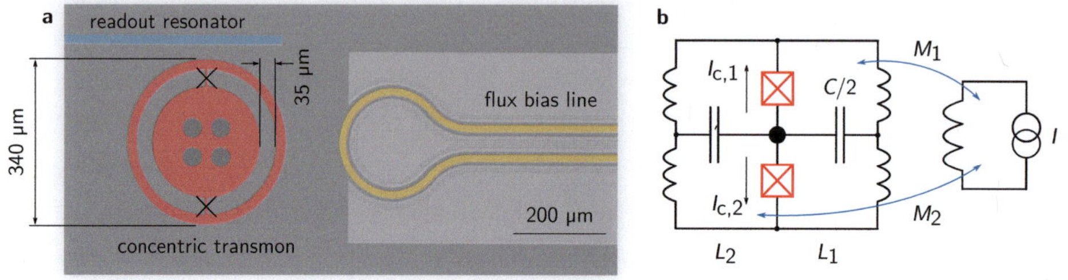

Figure 5.1: Architecture of the concentric transmon qubit (a) Optical micrograph of the fabricated sample. The dark grey colour corresponds to the intrinsic silicon substrate, while the coloured and brighter areas are covered with an Al film. The transmon consists of a central island with holes for flux trapping, that is surrounded by a concentric ring electrode. The electrodes are interconnected by two Josephson junctions, forming a gradiometric dc SQUID. The open end of a microstrip $\lambda/2$-resonator (coloured in blue) capacitively couples to the concentric transmon for dispersive qubit readout. An on-chip flux bias coil (yellow) allows for fast tuning of the qubit transition frequency since it couples asymmetrically to the two SQUID loops. It is designed in coplanar geometry in order to match its impedance to 50 Ω. The flux bias line is grounded at one of its ends on chip. A superconducting groundplane is applied to the backside of the substrate. (b) Simplified lumped-element circuit diagram. The black dot in the middle corresponds to the central qubit island. One can recognize two closed loops (L_1, L_2) sharing the two Josephson junctions. The large loop size in the geometry amounts to a geometric inductance contribution, as indicated in the circuit. Since the mutual inductances to the flux bias line are not equal, $M_1 \neq M_2$, the effective critical current of the SQUID can be tuned.

5.2 Design and architecture

The design and architecture of the concentric transmon qubit is depicted in Fig. 5.1(a). The two capacitor pads forming the large shunt capacitance of the transmon are implemented by a central disk island and a concentrically surrounding ring. The two islands are interconnected by two Josephson junctions forming a gradiometric SQUID. The transmon islands act as coplanar electrodes [San+13] giving rise to the total qubit capacitance $C = 81\,\mathrm{fF}$, including the contribution by the ground plane and coupling capacitances.

For readout and control purposes we embed the qubit in a microstrip resonator circuit, forming a familiar cQED system. The $\lambda/2$ readout resonator has a net capacitive coupling to the concentric transmon since it couples more strongly to the ring electrode than to the central disk. The coupling strength is in first order independent of the orientation of the concentric transmon due to the lumped-element nature of the qubit.

A $50\,\Omega$ impedance matched on-chip flux bias line located next to the qubit allows for fast flux tuning of the qubit frequency due to the imposed asymmetry. This guarantees high experimental flexibility and enables full tomographic control. The gradiometric flux loop renders the qubit insensitive to external uniform magnetic fields and thus suppresses external flux noise.

A simplified lumped-element circuit diagram of the concentric transmon is schematically depicted in Fig. 5.1(b). Considering loop L_1 as the primary transmon loop, the gradiometric dc-SQUID architecture can be recognized. The kinetic inductance of the superconducting aluminum wire can be neglected due to its finite thickness and the width of the Josephson junction leads. The effective critical current $I_{\mathrm{c,eff}}$ of the SQUID is tuned by applying an inhomogeneous magnetic field supplied by the on-chip flux bias line. From flux quantization in both closed loops L_1, L_2,

$$\phi_1 - \phi_2 = 2\pi\frac{\Phi_1}{\Phi_0}, \tag{5.1}$$

$$-\phi_1 + \phi_2 = 2\pi\frac{\Phi_2}{\Phi_0} \tag{5.2}$$

we obtain

$$\phi_1 - \phi_2 = \pi\frac{\Phi_1 - \Phi_2}{\Phi_0} = \pi\frac{\Delta\Phi}{\Phi_0}, \tag{5.3}$$

such that the effective critical current of the primary loop L_1 becomes

$$I_{\mathrm{c,eff}} = 2I_{\mathrm{c}}\left|\cos\left(\frac{\pi\Delta\Phi}{2\Phi_0}\right)\right| \tag{5.4}$$

and is therefore $2\Phi_0$-periodic in the flux asymmetry $\Delta\Phi = |\Phi_1 - \Phi_2|$ between the loops. $\Phi_0 = h/2e$ denotes the magnetic flux quantum. We analytically calculate the net mutual inductance to the flux bias line to be $2.3\,\mathrm{pH}$ by applying the double integral Neumann formula [Jac62]. The calculation yields a flux bias current of $1.8\,\mathrm{mA}$, required to induce $2\Phi_0$ in the primary transmon loop. This is confirmed in qubit spectroscopy shown in Fig. 5.3(b), where we observe a periodicity of $\sim 1.7\,\mathrm{mA}$.

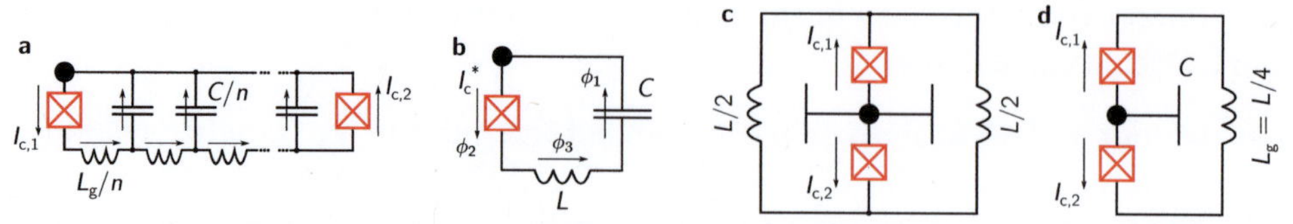

Figure 5.2: Schematic circuit diagrams for the concentric transmon qubit (a) Exact lumped-element representation of the qubit circuit accounting for the distribution of capacitance and inductance per unit length. (b) Simplified effective circuit diagram used for circuit quantization. The dc SQUID is treated as a single effective Josephson junction with critical current I_c^*, connected in series with an effective inductance L and the qubit capacitance C. It is suggestive to identify L with the inductance of the complete qubit ring electrode. (c) Geometric representation of the circuit. The qubit capacitance is centre symmetrically dispersed. (d) Diagram motivating the choice of effective inductance L. Two half rings with inductance $L/2$, respectively, are connected in parallel, forming the geometric inductance $L_g = L/4$.

The position of the flux sweet spot with maximum transmon frequency is dependent on the number of trapped flux quanta in the two SQUID loops. The dispersion maximum appears at zero externally applied flux, $\Delta\Phi = 0$, if the difference in number of trapped fluxes in the two loops, respectively, is an even number. This corresponds to the familiar scenario for conventional tunable transmon qubits. If the difference in trapped flux quanta is an odd number when cooling through T_c, the dispersion is shifted by Φ_0 and has its minimum frequency at $\Delta\Phi = 0$. We confirmed this behaviour by more than ten cool-downs where we found the dispersion located symmetrically either around its maximum or its minimum frequency. A further small horizontal offset of the qubit spectrum is due to asymmetric flux trapping in the electrode areas.

We operate the device at a fundamental qubit transition frequency $\epsilon/2\pi = 6.85\,\text{GHz}$, far detuned from the readout resonator at $\omega/2\pi = 8.79\,\text{GHz}$ in order to reduce Purcell dissipation. The Josephson energy $E_J/h = 29\,\text{GHz}$ dominates the charging energy $E_C/h = 0.24\,\text{GHz}$, rendering a $E_J/E_C = 120$ well within the transmon regime. This assures the circuit to be inherently insensitive to charge noise and features an intrinsic self-biasing [Koc+07]. For the operation point we extract a relative anharmonicity of $\alpha_r = -3.4\%$.

5.3 Circuit quantization

The geometric inductance of the large loops in the concentric transmon architecture constitutes a notable contribution to the total qubit inductance. To account for this, we derive a generalized system Hamiltonian to model our circuit. An exact lumped element representation of the qubit circuit is depicted in Fig. 5.2(a). The qubit capacitance is distributed centre symmetrically into n small parallel capacitances C/n and the total inductance L_g is split into a series of n inductors of inductance L_g/n. In order to simplify the analytic treatment, we approximate this exact circuit by the effective simplified circuit diagram depicted in Fig. 5.2(b), consisting of a single capacitor of capacitance C connected in series with an effective inductance L and a single Josephson junction of effective critical current I_c^*. The total critical current of the SQUID is treated as an effective parameter $I_c^* = 1.5 I_c$ to account for model inaccuracies. The geometric inductance L_g we extract in this analysis is related to the effective inductance L via $L_g = L/4$. This can be motivated since L_g is formed by two half rings in parallel, see Fig. 5.2(c), (d). This semi-quantitative treatment is justified by a good agreement of the presented model with measured data and simulation results. In the following we present a detailed derivation of the system Hamiltonian based on the simplified circuit diagram depicted in Fig. 5.2(b).

From current conservation at the two independent active nodes we get

$$\frac{\hbar}{2e} C\ddot{\phi}_1 = I_c^* \sin\phi_2 = \frac{\hbar}{2e}\frac{1}{L}\phi_3. \tag{5.5}$$

The phases ϕ_i in Fig. 5.2(b) correspond to the respective voltage drop across each of the components of the circuit according to

$$\frac{\hbar}{2e}\phi_i = \int_{t_0}^{t} \mathrm{d}t' U_i(t'). \tag{5.6}$$

The directed voltage drops in a closed network add up to zero according to Kirchhoff's rule. In a closed loop, the integration constant in Eq. (5.6) is the total magnetic flux penetrating the loop. Since the network in Fig. 5.2(b) is an open loop in the sense of the flux quantization law, we can write

$$\phi_1 + \phi_2 + \phi_3 = 0. \tag{5.7}$$

We eliminate ϕ_2 using Eq. (5.5) and insert in Eq. (5.7)

$$\phi_3 + \arcsin(c\phi_3) = -\phi_1, \tag{5.8}$$

introducing the notation

$$c = \frac{\hbar}{2e}\frac{1}{I_c^* L}. \tag{5.9}$$

An approximate solution to Eq. (5.8) can be written as

$$\phi_3 = -\frac{1}{(1+c)^2}\phi_1 - \frac{c}{(1+c)^2}\sin\phi_1 \tag{5.10}$$

which is exact when $\phi_1 \ll 1$ or $c \gg 1$. Plugging into Eq. (5.5) and writing $\phi_1 \equiv \phi$, we reduce the set of equations to a single equation of motion,

$$\frac{\hbar}{2e} C\ddot{\phi} = -\frac{\hbar}{2e}\frac{1}{L}\frac{1}{(1+c)^2}\phi - \frac{\hbar}{2e}\frac{1}{L}\frac{c}{(1+c)^2}\sin\phi. \tag{5.11}$$

With the Euler-Lagrange equation

$$\frac{\mathrm{d}}{\mathrm{d}t}\frac{\partial \mathcal{L}}{\partial\left(\frac{\hbar}{2e}\dot{\phi}\right)} = \frac{\partial \mathcal{L}}{\partial\left(\frac{\hbar}{2e}\phi\right)} \tag{5.12}$$

we obtain the Lagrange function

$$\mathcal{L} = \frac{C}{2}\left(\frac{\hbar}{2e}\dot{\phi}\right)^2 - \frac{1}{2L}\frac{1}{(1+c)^2}\left(\frac{\hbar}{2e}\phi\right)^2 + \left(\frac{\hbar}{2e}\right)^2\frac{1}{L}\frac{c}{(1+c)^2}\cos\phi. \tag{5.13}$$

Employing the definitions of the Josephson energy $E_\mathrm{J} = \hbar I_\mathrm{c}/2e$, and the inductive energy $E_\mathrm{L} = (\hbar/2e)^2(2L_\mathrm{g})^{-1}$, the parameter c becomes

$$c = \frac{2e}{\hbar}\frac{1}{1.5I_\mathrm{c}}\left(\frac{\hbar}{2e}\right)^2\frac{2}{8L_\mathrm{g}} = \frac{E_\mathrm{L}}{3E_\mathrm{J}}. \tag{5.14}$$

Using the conjugated variable for the charge number $N = C\frac{\hbar}{(2e)^2}\dot{\phi}$, we arrive at the Hamiltonian

$$\hat{H} = 4E_\mathrm{C}(\hat{N} - n_\mathrm{g})^2 - E_\mathrm{J}\frac{6E_\mathrm{L}^2}{(6E_\mathrm{J} + 2E_\mathrm{L})^2}\cos\hat{\phi} + E_\mathrm{L}\frac{9E_\mathrm{J}^2}{(6E_\mathrm{J} + 2E_\mathrm{L})^2}\hat{\phi}^2 \tag{5.15}$$

with charging energy $E_\mathrm{C} = e^2/2C$.

The constructed system Hamiltonian reads

$$\hat{H} = 4E_\mathrm{C}(\hat{N} - n_\mathrm{g})^2 - \tilde{E}_\mathrm{J}\cos\hat{\phi} + \tilde{E}_\mathrm{L}\hat{\phi}^2, \tag{5.16}$$

introducing the quantities

$$\tilde{E}_\mathrm{J} = E_\mathrm{J}\frac{6E_\mathrm{L}^2}{(6E_\mathrm{J} + 2E_\mathrm{L})^2}, \quad \tilde{E}_\mathrm{L} = E_\mathrm{L}\frac{9E_\mathrm{J}^2}{(6E_\mathrm{J} + 2E_\mathrm{L})^2}. \tag{5.17}$$

Note that $\tilde{E}_\mathrm{L} = 0$ in the limit $E_\mathrm{L} \to \infty$, such that the original transmon Hanmiltonian is recovered up to a factor of 3/2 in the Josephson term that comes from the effectiveness of the critical current in the model.

Expanding the cosine in Eq. (5.16) up to fourth order in $\hat{\phi}$ yields

$$\hat{H} = 4E_\mathrm{C}(\hat{N} - n_\mathrm{g})^2 + \left(\frac{\tilde{E}_\mathrm{J}}{2} + \tilde{E}_\mathrm{L}\right)\hat{\phi}^2 - \frac{\tilde{E}_\mathrm{J}}{4!}\hat{\phi}^4 + \mathrm{const.} \tag{5.18}$$

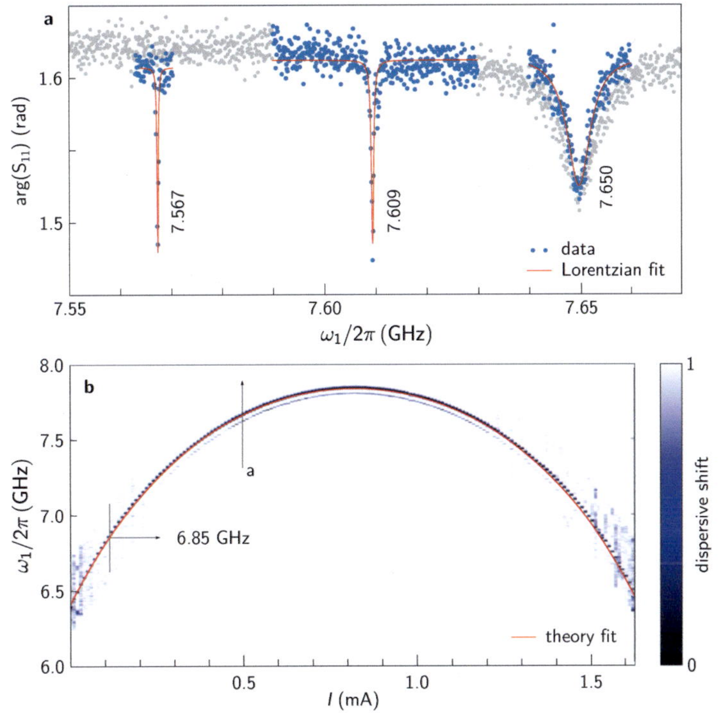

Figure 5.3: Concentric transmon spectroscopy (a) Fundamental qubit transition and lowest order multi-photon transitions. From the extracted transition frequencies $\epsilon_{01}/2\pi = 7.6496\,\mathrm{GHz}$, $\frac{1}{2}\epsilon_{02}/2\pi = 7.6094\,\mathrm{GHz}$ and $\frac{1}{3}\epsilon_{03}/2\pi = 7.5673\,\mathrm{GHz}$ we obtain three equations using Eq. (5.21). The transition peaks were recorded at different excitation powers (blue data points) and fitted to Lorentzians. (b) Frequency dispersion of the concentric transmon qubit with a dc current I applied to the flux bias line. The thick blue line denotes the measured qubit transition frequency ϵ_{01} which is fitted to our theory model (red). Good agreement is obtained for taking into account a geometric inductance contribution. The faint blue line visible close to the flux sweet spot shows the two-photon transition $\frac{1}{2}\epsilon_{02}$. We performed a normalization of each individual column, sich that a weaker signal close to the points of minimum frequency leads to a enhancement of relative background noise. Qubit measurements presented in Sec. 5.5, 5.6 are taken at a qubit frequency $\epsilon_{01}/2\pi = 6.85\,\mathrm{GHz}$.

We identify $4E_C \equiv \hbar^2/2m$ and $\tilde{E}_J/2 + \tilde{E}_L \equiv m\omega^2/2$ to cast the Hamiltonian in Eq. (5.18) into the standard form of the harmonic oscillator for terms of order $O(\hat{\varphi}^2)$. This yields the phase operator

$$\hat{\varphi} = \left(\frac{E_C}{\tilde{E}_J/2 + \tilde{E}_L}\right)^{1/4} (\hat{a}^\dagger + \hat{a}) \tag{5.19}$$

in the representation of creation (annihilation) operators $\hat{a}^\dagger$ ($\hat{a}$). Plugging into Eq. (5.18) and taking into account the bosonic commutation relation, $[\hat{a}, \hat{a}^\dagger] = 1$, yields

$$\hat{H} = 4\sqrt{E_C(\tilde{E}_J/2 + \tilde{E}_L)}\,\hat{a}^\dagger\hat{a} - \frac{\tilde{E}_J E_C}{4(\tilde{E}_J/2 + \tilde{E}_L)}\left((\hat{a}^\dagger\hat{a})^2 + \hat{a}^\dagger\hat{a}\right) + \text{const.} \tag{5.20}$$

and we can find the energies E_{0j} of concentric transmon levels $|j\rangle$ relative to the ground state to be

$$E_{0j} = 4\sqrt{E_C(\tilde{E}_J/2 + \tilde{E}_L)}j - \frac{\tilde{E}_J E_C}{4(\tilde{E}_J/2 + \tilde{E}_L)}(j^2 + j). \tag{5.21}$$

For the concentric transmon under investigation, Fig. 5.3(a) shows the fundamental qubit transition ϵ_{01} and the lowest order multi-photon transitions $1/2\epsilon_{02}$ and $1/3\epsilon_{03}$, measured close to the current sweet spot of the spectrum, see the arrow in Fig. 5.3(b). Equating Eq. (5.21) for $j = 1, 2, 3$ and using fitted transition data as depicted in Fig. 5.3(a) yields three equations for the transmon parameters E_C, E_J and E_L. In an iterative approach using this transition data and by fitting measured spectroscopy data, we can extract a set of parameters which are in good agreement with both data sets. The spectrum fit is carried out using Eq. (5.21) evaluated for $j = 1$, see Fig. 5.3(b) and

$$E_J(\Delta\Phi) = \frac{\hbar}{2e} I_c^m \left|\cos\left(\pi\frac{\Delta\Phi}{\Phi_0}\right)\right| \sqrt{1 + d^2\tan^2\left(\pi\frac{\Delta\Phi}{\Phi_0}\right)}. \tag{5.22}$$

I_c^m denotes the total critical current of the concentric transmon at the flux sweet spot, where $\Delta\Phi = 0$ and we omitted a phase offset used for fitting purposes. The square root term in Eq. (5.22) accounts for a relative asymmetry between the Josephson junctions [Koc+07], specified by the asymmetry parameter $d = (I_{c,1} - I_{c,2})/(I_{c,1} + I_{c,2})$. A separate fit of the qubit spectrum indicates an asymmetry parameter of $d = 0.32$, causing a decrease in tunability range of the circuit. The minimum qubit frequency of the investigated sample is 6.3 GHz.

Table 5.1 summarizes the parameters extracted from the described fitting method based on measured data shown in Fig. 5.3. Microwave simulations indicate a shunt capacitance of $C_{sh} = 58$ fF, yielding a capacitance of ~ 20 fF between the disk electrode and ground, which contributes to the effective transmon capacitance. A separate geometric finite elements simulation yields $L_g \sim 0.2$ nH, deviating by a factor of three from the fitted value.

Table 5.1: Summary of extracted parameters of the investigated concentric transmon qubit Errors Δ in E_C and C are assumed, other errors are standard deviations as extracted from the fit. I_c denotes the total critical current of both Josephson junctions. The contribution of the geometric inductance relative to the total qubit inductance $L_g + L_j$ at 6.85 GHz is $L_g/(L_g + L_j) \sim 10\%$, where L_j is the kinetic inductance of the Josephson junctions.

	E_J	I_c	E_C	C	E_L	L_g
sweet spot	45 GHz	90 nA	0.24 GHz	81 fF	128 GHz	0.64 nH
6.85 GHz	29 GHz	58 nA				
Δ	±12 GHz	±23 nA	±0.03 GHz	±10 fF	±30 GHz	±0.16 nH

5.4 Fabrication

The fabrication process consists of two subsequent deposition and lithography steps. The feedline, the microstrip resonator and the flux bias line are structured by optical lithography in a lift-off process of a 50 nm thick Al film. In a successive step, the concentric transmon including the Josephson junctions are patterned using electron beam lithography. The Josephson junctions are formed by shadow angle evaporation as described in Sec. 4.1. A 50 nm thick Al film is applied on the backside of the intrinsic silicon substrate.

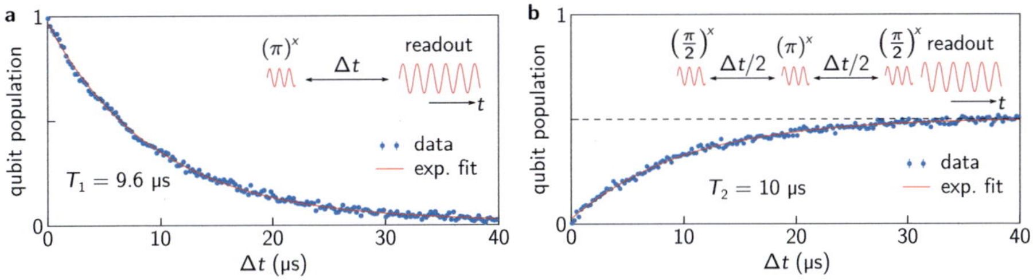

Figure 5.4: Measurement of the dissipative dynamics (a) By exciting the qubit with a previously calibrated π pulse and reading out the qubit state after a varying time Δt, we record a relaxation time $T_1 = 9.6$ μs. Relaxation times up to ~ 15 μs were measured for later sample generations. (b) Echo dephasing time $T_2 = 10$ μs, measured with a Ramsey pulse sequence and one additional Hahn echo, as shown in the inset. The qubit operation frequency was $\epsilon/2\pi = 6.85$ GHz, corresponding to a detuning of 1.9 GHz below the readout resonator frequency.

5.5 Dissipative dynamics

The dissipative dynamics of the investigated concentric transmon is depicted in Fig. 5.4. At a qubit operation frequency of $\epsilon/2\pi = 6.85\,\mathrm{GHz}$ we extract $T_1 = 9.6\,\mu\mathrm{s}$ by exciting the qubit with a previously calibrated π pulse and measuring its population after varying times Δt. We engineered the concentric transmon to have a reduced sensitivity to its major loss channels, namely spontaneous Purcell emission, dielectric loss, and radiative decay. Losses due to quasi-particle tunneling processes typically impose a T_1 limitation at around $\sim 1\,\mathrm{ms}$ [Ris+13], having no considerable effect on the lifetime of our circuit.

In the following, we provide a loss participation ratio analysis in order to explain the measured value for T_1. The results are summarized in Tab. 5.2.

Table 5.2: Calculated loss contributions fo the concentric transmon qubit The main contribution $\Gamma^{-1}_{1,\mathrm{ind}}$ arises from inductive coupling to the flux bias line, leading to a total Purcell limitation of $\Gamma^{-1}_{1,P} = 16\,\mu\mathrm{s}$. The estimated reciprocal sum Γ^{-1}_{Σ} is in good agreement with the measured value for T_1.

	Purcell		Defects	Radiation	Reciprocal sum
$\Gamma^{-1}_{1,\mathrm{sm}}$	$\Gamma^{-1}_{1,\mathrm{ind}}$	$\Gamma^{-1}_{1,\mathrm{cap}}$	$\Gamma^{-1}_{1,\mathrm{TLF}}$	$\Gamma^{-1}_{1,\mathrm{rad}}$	Γ^{-1}_{Σ}
47 µs	32 µs	$\sim$ 87 µs			
	$\Gamma^{-1}_{1,\mathrm{P}} = 16\,\mu\mathrm{s}$		$\sim$ 26 µs	$\gtrsim$ 100 µs	8.9 µs

5.5.1 Purcell decay

We find a coupling limited decay rate $\kappa = 26 \times 10^6\,\mathrm{s}^{-1}$, corresponding to a line width of 4.1 MHz of the dispersive readout resonator, which is close to the design value. The qubit lifetime is potentially Purcell limited by spontaneous emission into modes that are nearby in frequency. Major contributions are the dispersive single-mode decay into the capacitively coupled readout resonator as well as emission into the flux bias line [Koc+07; Hou+08] due to inductive coupling. The coupling strength $g/2\pi = 55\,\mathrm{MHz}$ between qubit and resonator is extracted from the dispersive shift of the resonator [Bra+15].

The single-mode Purcell limitation can be found from Fermi's Golden rule using the dipolar coupling Hamiltonian

$$\hat{H}_\mathrm{c} = \hbar g \hat{\sigma}_x \left(\hat{b}^\dagger + \hat{b}\right) \tag{5.23}$$

between qubit and resonator with $\hat{b}^\dagger$ ($\hat{b}$) bosonic creation (annihilation) operators in the harmonic Fock space of the resonator. For the decay rate Γ_1 from the excited qubit state $|\mathrm{e}\rangle$ to its groundstate $|\mathrm{g}\rangle$ under photon creation in the resonator $|n\rangle \rightarrow |n+1\rangle$ we obtain

$$\Gamma_1 = \frac{2\pi}{\hbar} \sum_i \hbar^2 g^2 |\langle 0| \otimes \langle n+1|\hat{\sigma}_x \left(\hat{b}^\dagger + \hat{b}\right) |1\rangle \otimes |n\rangle|^2 \frac{\delta(\epsilon - \omega_i)}{\hbar}, \tag{5.24}$$

with discrete contributions $\delta(\epsilon - \omega_i)$ from the Dirac function. Assuming small temperatures, the resonator is in its groundstate and we can set $n = 0$. Writing the sum in Eq. (5.24) as an integral introduces the spectral density function $\rho(\omega, \omega_r)$ of the resonator mode. Applying a normalized Lorentzian shape

$$\rho(\omega', \omega) = \frac{1}{\pi} \frac{\kappa/2}{(\kappa/2)^2 + (\omega' - \omega)^2}, \quad \frac{1}{\pi} \int_{-\infty}^{\infty} \mathrm{d}\omega' \frac{\kappa/2}{(\kappa/2)^2 + \omega'^2} = 1 \tag{5.25}$$

with full width at half maximum κ yields

$$\Gamma_1 = 2\pi g^2 \int \mathrm{d}\omega' \rho(\omega', \omega) \delta(\epsilon - \omega') = 2g^2 \frac{\kappa/2}{(\kappa/2)^2 + (\epsilon - \omega)^2} = 2g^2 \frac{\kappa/2}{(\kappa/2)^2 + \Delta^2}, \tag{5.26}$$

where $\Delta = \epsilon - \omega$ is the detuning between qubit and resonator. In the dispersive limit $\Delta \gg \kappa$, we calculate a single-mode Purcell limitation induced by the readout resonator of $\Gamma_{\Gamma 1,sm} = \kappa\,(g/\Delta)^2 = 1/47\,(\mu\mathrm{s})^{-1}$ at a detuning $\Delta = 1.94\,\mathrm{GHz}$. While the formula provided in Eq. (5.26) is exact in the dispersive limit, it can reproduce the expected envelope of the decay rate on resonance, where $\Gamma_1(\Delta = 0) \propto g^2/\kappa$.

For a qubit being capacitively coupled to an environment of impedance $Z_0(\omega)$ via a capacitance C_c, the total impedance seen by the qubit becomes

$$Z(\omega) = -\frac{i}{\omega C_c} + Z_0(\omega). \tag{5.27}$$

The decay rate Γ_1 due to classical zero-point noise [Hou+08; Mar+03] can be written as

$$\Gamma_Z = \frac{\Re Y(\epsilon)}{C}, \tag{5.28}$$

where C is the total qubit capacitance and $Y(\epsilon)$ denotes the admittance of the environment at the qubit frequency ϵ. Since the effective qubit resistance takes the form $R = 1/\Re Y(\epsilon)$, the decay rate in Eq. (5.28) is the inverse time constant of the RC element. Using Eq. (5.27) we can write

$$\Re Y(\epsilon) = \Re 1/Z(\epsilon) = \frac{\Re Z(\epsilon)}{\Re Z(\epsilon)^2 + \Im Z(\epsilon)^2} = \frac{Z_0(\epsilon)}{Z_0(\epsilon)^2 + 1/(\epsilon C_c)^2} = Z_0(\epsilon)\epsilon^2 C_c^2. \tag{5.29}$$

The last step in Eq. (5.29) is valid for $1/(\epsilon C_c) \ll Z_0(\epsilon)$. Substituting Eq. (5.29) into Eq. (5.28) yields the multi-mode decay rate

$$\Gamma_Z = \frac{Z_0 \epsilon^2 C_c^2}{C}. \tag{5.30}$$

In analogy to this result, one can write down the decay rate Γ for a qubit being inductively coupled to an arbitrary impedance:

$$\Gamma_Z = \frac{\epsilon^2 M^2}{Z_0 L_\Sigma}, \tag{5.31}$$

with M the mutual inductance between qubit and impedance load and L_Σ the total qubit inductance.

The multi-mode Purcell limitation due to inductive coupling of the concentric transmon to the flux bias line of impedance $Z_0 = 50\,\Omega$ is $\Gamma_{1,\mathrm{ind}} = 1/32\,(\mu\mathrm{s})^{-1}$, with mutual inductance $M = 2.3\,\mathrm{pH}$. The total qubit inductance $L_\Sigma = 6.3\,\mathrm{nH}$ includes the kinetic inductance of the Josephson junctions and the geometric inductance contribution.

In analogy, we find a Purcell decay rate due to capacitive coupling (~ 0.1 fF) to the flux bias line, we estimate $\Gamma_{1,\mathrm{cap}} \sim 1/87\,(\mu\mathrm{s})^{-1}$. The presented values are approximations and give a rough estimate of the expected Purcell loss. A more stringent analysis would require a full 3D electromagnetic simulation of the circuit for instance by the blackbox quantization technique [Nig+12; SAD14].

5.5.2 Dielectric loss

The coplanar concentric transmon is embedded in a microstrip geometry, where the ground reference on the device side of the substrate is substituted by a backside metallization. The largest fraction of the electric field energy is stored in the substrate, and the field strength at incoherent and weakly coupled defects residing in surface and interface oxides of the sample is reduced due to an increased mode volume. Highest fields in the geometry appear in the gap between the centre island and the ring of the concentric transmon within the substrate. The smallest dimension in the geometry is given by the electrode spacing of 35 µm, see Fig. 5.1, which roughly corresponds to the equivalent substrate thickness in vacuum of 380 µm/$\epsilon_r \sim$ 33 µm. While electric fields between the transmon electrodes are enhanced by the dielectric substrate, we provide a microscopic analysis of the defect loss based on the dimensions in our geometry. The estimate is demonstrated to be in good agreement with a finite element simulation.

Microscopic treatment

From the vacuum energy of the transmon we extract a weighted mean electrical field strength $|\vec{E}| = 2.3\,\mathrm{Vm}^{-1}$ in the surface and interface oxide of an estimated effective volume of $V = 50 \times 10^{-18}\,\mathrm{m}^3$, compare Fig. 5.5. We assume a maximum dipole moment $|\vec{d}_0| = 1.6\,\mathrm{e}$ of dielectric defects, reported in literature [Mar+05; Col+10; MCL17] as the highest dipole moment observed in Josephson junction barriers and therefore yielding a worst case estimation. The maximum coupling strength to a single TLF therewith is $|\vec{E}||\vec{d}_0| \sim 90\,\mathrm{kHz}$. We employ a normalized dipole moment distribution [Mar+05] $P(p) = A\sqrt{1-p^2}/p$, with relative dipole moment $p = |\vec{d}|/|\vec{d}_0|$. Taking into account a normalized defect probability distribution [Mül+15] $P(\omega, \theta) = B\omega^{\alpha}\cos^{\alpha}\theta/\sin\theta$ allows us to estimate the mean relaxation rate $\Gamma_{1,\mathrm{TLF}}$ due to a single incoherent two level fluctuator (TLF) with averaged parameters to be

$$\int_0^{d_0} \mathrm{d}pP(p)\frac{p^2|\vec{E}|^2}{\hbar^2}\int_0^{\omega_{\mathrm{TLF}}}\mathrm{d}\omega\int_0^{\pi/2}\mathrm{d}\theta P(\omega,\theta)S(\epsilon). \tag{5.32}$$

ω denotes the TLF frequency that we integrate to a maximum of $\omega_{\mathrm{TLF}}/2\pi = 15\,\mathrm{GHz}$, and the angle θ accounts for the dipole coupling matrix element. The spectral function $S(\omega_q) = \sin^2\theta\,\gamma_2/(\gamma_2^2 + (\omega-\omega_q)^2)$ is essentially the Fourier transform of the coupling correlation function [Mül+15], with an assumed TLF dephasing rate $\gamma_2/2\pi = 10\,\mathrm{MHz}$. The averaged rate induced by a single TLF, given in Eq. (5.32), is multiplied by the number $N = \rho_0 V\hbar\omega_{\mathrm{TLF}}$ of defect TLF interacting with the qubit. With a distribution parameter $\alpha \sim 0.3$ [FI15] and a constructed defect density $\rho_0 = 4 \times 10^2\,(\mu\mathrm{m})^{-3}(\mathrm{GHz})^{-}1$ we compute a T_1 limitation due to the quasi-static bath of incoherent TLF to be $\Gamma_{1,\mathrm{TLF}} \sim 1/26\,(\mu\mathrm{s})^{-1}$. The choice of ρ_0 is consistent with literature [Bar+13; Mar+05] and is justified by a good agreement with a loss participation ratio analysis carried out via a finite element simulation, see the next paragraph. The calculated decay rate shows a very weak dependence on the employed parameters γ_2, α and integration cut-offs. Due to the small size of the Josephson junction, the defect density in the Josephson barrier itself is discrete and therefore highly reduced [Wei+11a; Wei+11b].

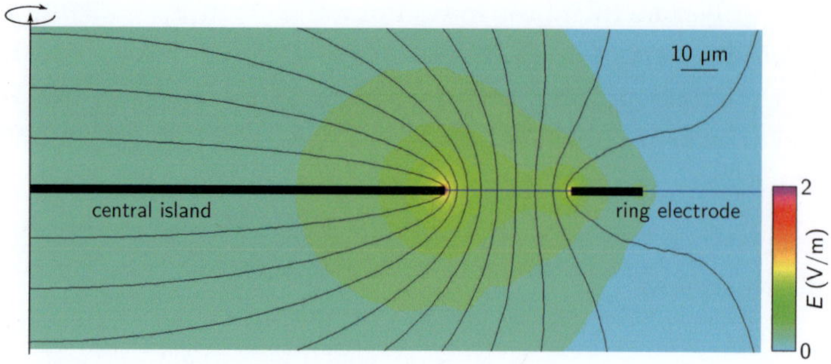

Figure 5.5: Finite element simulation of defect loss The schematic shows a side view of the concentric transmon qubit. The rotational symmetry axis is in the centre of the central transmon island and indicated with the arrow to the left. The electric field distribution is depicted in colours. Maximum values occur at the edges of the transmon electrodes and exceed the maximum value of the scale bar, such that these results are in reasonable agreement with the stated value for the mean electrical field strength in the oxide volume of $|\vec{E}| = 2.3\,\mathrm{Vm}^{-1}$.

Finite element simulation of defect loss

To validate the above microscopic analysis, we conduct a finite element simulation of the concentric transmon geometry [1]. The simulation layout with the electric field distribution is depicted in Fig. 5.5. We model the sample by assuming an oxide layer thickness of 3 nm, formed by AlO_x ($\epsilon_r = 11.5$), surrounding the central island and the concentric ring electrode of the qubit. For the interface between substrate and aluminum we assume an effective dielectric constant of $\epsilon_r = 6$, accounting for the contribution of SiO_2 ($\epsilon_r = 4$) in the interface oxide. The simulation yields an electric field energy fraction of 2.8×10^{-4} residing in the oxide volume. With a phenomenological loss tangent [PO10] $\delta_{AlO_x} = 3 \cdot 10^{-3}$ of AlO_x, this leads to a total effective loss tangent $\delta_{TLF} = 8.4 \times 10^{-7}$ which we attribute to dissipation by surface and interface oxide defects. Due to uncertainties in δ_{AlO_x}, the oxide layer thickness and ϵ_r, this value is to be considered exact within a factor of two. δ_{TLF} can be calculated to limit T_1 at $\Gamma_{1,TLF}^{-1} = 28\,\mu s$ which is in good agreement with the microscopically extracted value.

In the same spirit we can extract a relative electric field energy ratio in the silicon substrate of 0.92. With an intrinsic silicon loss tangent $\delta_{Si} < 1 \times 10^{-7}$, dissipation in the substrate can be neglected. This is confirmed by very high quality factors measured for TiN resonators on intrinsic silicon [San+12].

[1] Finite Element Method Magnetics (FEMM), version 4.2

5.5.3 Radiation loss

As pointed out in Ref. [San+13], radiative loss into the medium far field becomes apparent for qubits with a large electric dipole moment. In conventional pad geometries, radiative decay is reduced since the dipole of the mirror image, induced by the ground plane of the microstrip geometry, radiates in anti-phase, leading to destructive interference. The circular geometry investigated here brings about an additional decrease in radiated power by a strongly reduced electric dipole moment of the concentric electrodes. We analyse the effect by simulating the dissipated power in a conductive plane placed 1.5 mm above our geometry in the medium far field and compare the result to a conventional pad architecture. The internal quality factor of the qubit eigenmode indicates a radiative contribution of $\Gamma_{1,\mathrm{rad}}^{-1} \gtrsim 100\,\mu\mathrm{s}$. The radiative dissipation of a comparable qubit in pad geometry exceeds this value by more than an order of magnitude.

As compared to Purcell loss, the radiative decay rate is a measure for the confinement of electric fields inside the geometry and can be understood as the coupling strength to spurious modes in the imperfect electromagnetic environment of the qubit. In our sample box with an engineered lower mode cut-off frequency of 12 GHz, loss may be induced by bond wires, glue or the PCB. In addition, reduced dipole radiation allows for a closer packaging of qubits that can be advantageous in complex chip layouts, and looses the requirements on the utilized sample box.

To verify the presented loss participation ratio analysis, T_1 was measured close to the flux sweet spot and the associated Purcell contribution is calculated to be 4.7 µs, reducing the overall dissipation estimate to $\Gamma_{\Sigma}^{-1} = 3.7\,\mu\mathrm{s}$, while the measured value was $T_1 = \Gamma_1^{-1} = 1.8\,\mu\mathrm{s}$ [Bra+16]. The enhanced reduction in T_1 at the sweet spot is not fully understood. A possible explanation is the disregard of losses related to the geometric inductance such as quasi particles or magnetic vortices.

In continuous lifetime measurements, we observe temporal fluctuations in T_1 down to $\sim 2\,\mu\mathrm{s}$ for the operation point in Fig. 5.4. We conjecture that this is attributed to the discrete dynamics of two-level fluctuators (TLF), located in the small oxide volume at the leads of the Josephson junctions, where the electric field strength is enhanced. The temporal fluctuations decreased in later generations of the concentric transmon, which was fabricated in a single lithography step. A possible reason is a decreased TLF density due to a cleaner substrate prior to deposition of the concentric transmon.

In Fig. 5.4(b), we present a measurement of the spin echo dephasing time $T_2 = 10\,\mu\mathrm{s}$ by making use of a Ramsey sequence with one additional Hahn echo pulse. The two

$\pi/2$ pulses rotate the qubit Bloch vector into the equatorial plane and back to the quantization axis of the qubit prior to state readout such that the final qubit state is sensitive to phase rotations that are acquired during the time Δt. The echo π pulse in the middle of the two $\pi/2$ pulses mirrors the Bloch sphere and therefore reverts the drift direction of the qubit Bloch vector. Therewith, the technique compensates low-frequency fluctuations in the qubit transition frequency and renders the qubit insensitive to noise below $\sim 1/\max(\Delta t) = 25\,\text{kHz}$. We measure a Ramsey dephasing time $T_2^* \sim 2\,\mu\text{s}$ without echo pulse. We want to point out that a rather high T_2 is achieved in spite of the large loop size in the geometry. The transmon was detuned by $\sim 1\,\text{GHz}$ from its flux sweet spot, such that the slope in the dispersion is finite and flux noise causes dephasing. Apart from noise from the current bias source, dephasing is presumably induced by local magnetic fluctuators that sit on the surface of the metallizations. We show in Ref. [Bra+16] that the Ramsey dephasing time $T_2^* = 2.2\,\mu\text{s}$ is slightly increased at the flux sweet spot, where $T_1 = 1.8\,\mu\text{s}$, indicating a pure dephasing time of $\tau \sim 6\,\mu\text{s}$ being increasingly relaxation limited. The Purcell contribution at the flux sweet spot is calculated to be $4.7\,\mu\text{s}$, reducing the overall dissipation estimate to $\sim 3.7\,\mu\text{s}$.

In a non-tunable titanium nitride (TiN) version of the concentric transmon with a single Josephson junction and an opening in the outer ring, we find lifetimes of $T_1 \sim 50\,\mu\text{s}$ and Ramsey dephasing times up to $T_2^* \sim 60\,\mu\text{s}$. Similar lifetimes have been reported for TiN transmon qubits in the conventional capacitor pad design [Cha+13]. We expect an increase in coherence times also for the tunable concentric design by implementing it in a TiN material system due to a suppression of defect loss in surface and interface oxides.

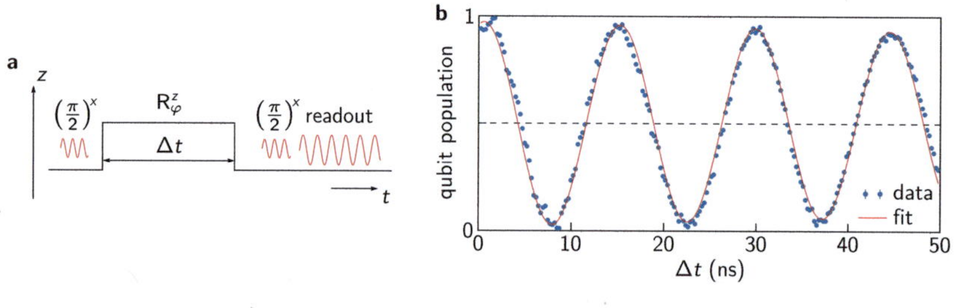

Figure 5.6: Demonstration of fast frequency tunability of the concentric transmon (a) Pulse sequence applied in the experiment. Between two projecting $\pi/2$ pulses we apply a $\hat{Z}$-rotation R_φ^z of amplitude $\eta = 32\,\mu\text{A}$ and length Δt. (b) The expectation value of the qubit state oscillates between its basis states $|g\rangle$, $|e\rangle$ with a relative Larmor frequency $\omega_L/2\pi = 65\,\text{MHz}$, according to the sinusoidal fit. A correction for the finite time constant in the bias tee was not performed as it can be neglected for such short pulse lengths.

The measured coherence times compare with other planar transmon geometries such as a non-tunable Al based transmon with decreased finger gap size [Cho+12] ($T_1 = 9.7\,\mu s$, $T_2^* = 10\,\mu s$) and the cross shaped transmon [Bar+13] ($T_1 = 40\,\mu s$, $T_2 = 20\,\mu s$), fabricated via molecular beam epitaxy deposition [Meg+12].

5.6 Frequency tunability

Figure 5.6 demonstrates fast frequency control of the concentric transmon. This is commonly referred to as $\hat{Z}$ control since Pauli's spin operator $\hat{\sigma}_z$ appears in the driving Hamiltonian. In the utilized pulse sequence shown in Fig. 5.6(a) we apply a $\hat{Z}$-rotation pulse of amplitude η in between a Ramsey sequence consisting of two projective $\pi/2$ pulses. We record the equatorial precession of the qubit Bloch vector at a relative Larmor frequency $\omega_L = 65\,\text{MHz}$ in Fig. 5.6(b). The $\hat{Z}$-pulse amplitude η translates into magnetic flux applied to the flux bias line. In the laboratory frame, this corresponds to a shift in qubit frequency by $\omega_L \propto \eta$, which is confirmed in a quasi-spectroscopic measurement by exciting the detuned qubit and measuring its transition frequency, see Ref. [Bra+16]. By increasing η, we demonstrated a fast frequency detuning of up to 200 MHz.

5.7 Coupling mechanisms

We propose the concentric transmon as a suitable candidate to establish a direct inductive longitudinal coupling between neighbouring qubits. This is referred to as $\hat{Z}$ coupling because the coupling Hamiltonian is proportional to $\hat{\sigma}_z$. Since the area of the magnetic flux loop is large, the magnetic dipole moment of our qubit is strongly enhanced as compared to conventional transmon designs where it is typically negligible. For two concentric transmon qubits separated by $\sim 50\,\mu m$, we estimate an inductive coupling in the range of $g_z/2\pi \sim 1\,\text{MHz}$ for an operation point where the flux dependent qubit spectrum has a large slope. This value is estimated by taking into account the full inductance matrix of a all four loops of a two-qubit unit cell [Ste16] and by applying the double integral Neumann formula [Jac62] to calculate individual mutual inductances M_{ij}. The technique considers the direction of currents flowing. We expect a longitudinal $\hat{Z}$ coupling in the range of few megahertz between two concentric transmon qubits sharing a fraction of their ring electrodes, which is currently explored in experiment.

At the given qubit distance, the conventional dipolar capacitive coupling, also referred to as a swap interaction, was simulated and experimentally confirmed to be 13 MHz [Ste16]. To further increase g_z, the mutual inductance in the two-qubit geometry must be increased. In a current experimental approach we galvanically connect the outer loops of two concentric transmon qubits in order to maximize M_{ij}. In an alternative approach, the mutual inductance of loops of adjacent qubits can be enhanced by arranging qubits on top of each other but transversally shifted in a flip chip approach [Ste16].

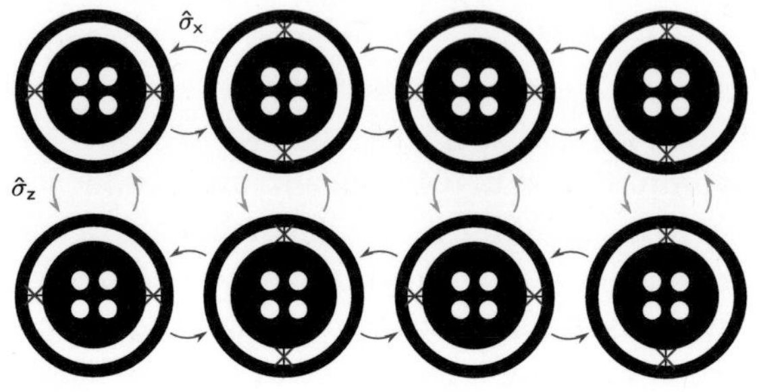

Figure 5.7: Possible implementation of the Fermi-Hubbard emulator with a double chain of concentric transmon qubits Within the two horizontal chains, qubits couple via a swap interaction proportional to $\hat{\sigma}_{x,j}\hat{\sigma}_{x,j+1}$. The linear chains form a ladder and adjacent qubits from different chains are coupled through a $\hat{z}$ type interaction proportional to $\hat{\sigma}_{z,j}\hat{\sigma}_{z,j}$. $\hat{z}$ interactions within chains vanish due to a $\pi/2$-rotation of neighbouring concentric transmons, while the swap interaction between the ladders can be suppressed by a mutual frequency detuning.

For a mutual frequency detuning of neighbouring qubits, the dipolar coupling can effectively suppressed, while the longitudinal coupling strength g_z does not depend on the qubit frequencies. The longitudinal coupling vanishes completely for adjacent qubits in plane arranged at a relative rotation angle of $\pi/2$, such that the concentric transmon geometry allows for a site-selective $\hat{Z}$ coupling. Figure 5.7 shows a schematic implementation of a quantum simulator of the Fermi-Hubbard model [Rei+16]. It consists of two linear chains of concentric transmons that are dipole coupled but with the $\hat{Z}$ coupling switched off along the two chains. This is achieved by alternating rotation angles and using the symmetry properties of the $\hat{Z}$ coupling. Adjacent qubits belonging to different chains are $\hat{Z}$ coupled while the dipole coupling is suppressed by a strong mutual frequency detuning.

The $\hat{Z}$ coupling scheme is highly promising for the field of quantum simulation in general, for instance in the context of implementing Ising spin models. Patterning an array of concentric transmon qubits featuring dipolar and $\hat{Z}$ coupling along two orthogonal directions while exploiting the strongly reduced off-site inductive interaction may be a route to implement a quantum neural network, which is a powerful tool in quantum computation [SSP14].

5.8 Conclusion

We have designed, fabricated and experimentally investigated a planar tunable qubit based on superconducting circuits that we call concentric transmon. With measured lifetimes and coherence times of up to 15 µs, the qubit design is competitive with conventional transmon geometries. The main limitations of the qubit lifetime are the Purcell effect induced by the readout resonator and the flux bias line, as well as loss from dielectric defects. Radiative loss was demonstrated to be suppressed due to the small electric dipole moment. A major advantage of our approach is the straightforward Al evaporation based fabrication process. The strong decrease in qubit lifetime due to defect loss in AlO_x is circumvented by a large mode volume in particular in the gap region between the transmon electrodes. We demonstrated full tomographic control of our quantum circuit and discuss the high potential of the presented qubit design for the implementation of a direct site-selective $\hat{Z}$ coupling between neighbouring qubits, rendering it a flexible building block for quantum simulation experiments.

6 The quantum Rabi model at ultra-strong coupling

The quantum Rabi model describes the fundamental mechanism of light-matter interaction. It consists of a two-level atom or qubit coupled to a quantized harmonic oscillator mode via a transversal interaction. In the small coupling regime, a rotating wave approximation can be applied and it reduces to the well-known Jaynes-Cummings model. The rotating wave approximation breaks down in the ultra-strong coupling regime, where the effective coupling strength g is comparable to the energy ω of the bosonic mode, and remarkable features in the system dynamics are revealed. We demonstrate an analog quantum simulation of an effective quantum Rabi model in the ultra-strong coupling regime, achieving a relative coupling ratio of $g/\omega \sim 0.6$. The quantum hardware of the simulator is a superconducting circuit embedded in a cavity quantum electrodynamics (cQED) setup. We observe fast and periodic quantum state collapses and revivals of the initial qubit state, being the most distinct signature of the synthesized model. [Bra+17]

6.1 Motivation

The quantum Rabi model in quantum optics describes the interaction between a two-level atom and a single quantized harmonic oscillator mode [Rab36; Rab37]. In the small coupling regime, which may still be strong in the sense of quantum electrodynamics (QED), a rotating wave approximation (RWA) can be applied and the quantum Rabi model reduces to the Jaynes-Cummings model [JC63], which captures most relevant scenarios in cavity and circuit QED. In the ultra-strong coupling (USC) and deep strong coupling regimes, where the coupling strength is comparable to or even exceeds the subsystem energies [Cas+10], the counter rotating terms in the interaction Hamiltonian can no longer be neglected and the RWA breaks down. As a consequence, the total excitation number is not conserved in the quantum Rabi model. Except for one recent paradigm of finding an exact solution [Bra11], an analytically closed solution of the quantum Rabi model does not exist due to the lack of a second conserved quantity, which renders it non-integrable. It is therefore appealing to investigate the non-classical features that emerge in the

USC regime and beyond in the context of quantum simulations. The dynamics of the quantum Rabi model is well reproduced by standard master equation approaches, such that it provides an ideal test bed for experimental simulation techniques.

Apart from its intrinsic quantumness, rising interest in the Rabi model is inspired by strong advances of experimental capabilities [Cas+10; AN10; BGB11; AR16]. The specific spectral features of the USC regime and the consequent breakdown of the RWA were previously observed with a superconducting circuit by implementing an increased physical coupling strength [Nie+10; For+10; For+16b]. A similar approach involving a flux qubit coupled to a single-mode resonator allowed to access the deep strong coupling regime in a closed system [Yos+16]. The USC regime was reached before by dynamically modulating the flux bias of a superconducting qubit, reaching a coupling strength of about 0.1 of the effective resonator frequency [Li+13].

In our approach, we create an effective quantum Rabi Hamiltonian in the USC regime by applying an analog quantum simulation scheme based on the application of microwave Rabi drive tones. By a decrease of the subsystem energies, the USC condition is satisfied in the effective rotating frame, which allows us to observe the distinct model dynamics. The scheme may be a route to efficiently generate non-classical cavity states [Leg+15; Kir+13; Vla+13], see Sec. 6.9, and may be extended to explore relevant physical models such as the Dirac equation in (1+1) dimensions. Its characteristic dynamics is expected to display a *Zitterbewegung* in the spacial quadrature of the bosonic mode [Bal+12]. This dynamics has been observed with trapped ions [Ger+10], likewise based on a Hamiltonian that is closely related to the USC Rabi model.

It has been shown recently that a quantum phase transition, typically requiring a continuum of modes, can appear already in the quantum Rabi model under appropriate conditions [HPP15]. The experimental challenge is projected onto the coupling requirements in the model which may be accomplished with the simulation scheme presented. This can be a starting point to experimentally investigate critical phenomena in a small and well-controlled quantum system [HP16]. By using a digital simulation approach, the dynamics of the quantum Rabi model at USC conditions was similarly studied very recently [Lan+17].

In our experiment we first simulate the USC quantum Rabi model in the absence of a qubit energy term, achieving a relative coupling strength of up to 0.6. This is achieved by applying a single Rabi drive. We observe periodically recurring quantum state collapses and revivals in the qubit dynamics as one of the characteristic signatures of USC. The collapse-revival dynamics appears most clearly in the absence of the qubit energy term in the quantum Rabi Hamiltonian. In a second step, we use our device to simulate the full quantum Rabi model and are able to observe the onset of an additional substructure in the qubit time evolution. With this proof of principle experiment we validate the experimental feasibility of the proposed analog quantum simulation scheme and demonstrate the potential of superconducting circuits for the field of quantum simulation.

6.2 Simulation Scheme

The quantum Rabi Hamiltonian reads

$$\frac{\hat{H}}{\hbar} = \frac{\epsilon}{2}\hat{\sigma}_z + \omega\hat{b}^\dagger\hat{b} + g\hat{\sigma}_x\left(\hat{b}^\dagger + \hat{b}\right), \tag{6.1}$$

with ϵ the qubit energy splitting, ω the bosonic mode frequency and g the transversal coupling strength. $\hat{\sigma}_i$ are Pauli matrices with $\hat{\sigma}_z\left|\mathrm{g}\right\rangle = -\left|\mathrm{g}\right\rangle$ and $\hat{\sigma}_z\left|\mathrm{e}\right\rangle = \left|\mathrm{e}\right\rangle$, where $\left|\mathrm{g}\right\rangle$, $\left|\mathrm{e}\right\rangle$ denote the eigenstates of the computational qubit basis. $\hat{b}^\dagger$ ($\hat{b}$) are creation (annihilation) operators in the Fock space of the bosonic mode. Both elements of the model are physically implemented in the experiment, with a small geometric coupling $g \ll \epsilon, \omega$. In order to access the USC regime, we follow the scheme proposed in Ref. [Bal+12]. It is based on the application of two transversal microwave Rabi drive tones coupling to the qubit. The effective USC quantum Rabi Hamiltonian is created in the frame rotating with the dominant drive frequency. In the engineered Hamiltonian, the effective mode energies are adjusted by the Rabi drive parameters.

The driven laboratory frame Hamiltonian takes the form

$$\begin{aligned}\frac{\hat{H}_\mathrm{d}}{\hbar} =& \frac{\epsilon}{2}\hat{\sigma}_z + \omega\hat{b}^\dagger\hat{b} + g\hat{\sigma}_x\left(\hat{b}^\dagger + \hat{b}\right) \\ &+ \hat{\sigma}_x\eta_1\cos(\omega_1 t + \varphi_1) + \hat{\sigma}_x\eta_2\cos(\omega_2 t + \varphi_2).\end{aligned} \tag{6.2}$$

η_i denote the amplitude and ω_i the frequency of drive i. φ_i is the relative phase of drive i in the coordinate system of the qubit Bloch sphere in the laboratory frame. Within the RWA where $\eta_i/\omega_i \ll 1$, the φ_i enter as relative phases of the transversal coupling operators $e^{-\mathrm{i}\varphi_i}\hat{\sigma}_+ + \mathrm{h.c.}$, where $\hat{\sigma}_\pm = 1/2\left(\hat{\sigma}_x \pm \mathrm{i}\hat{\sigma}_y\right)$ denote Pauli's ladder operators. In the following, we set $\varphi_i = 0$ to recover the familiar $\hat{\sigma}_x$ coupling without loss of generality.

By applying the unitary transformation

$$\hat{U} = \exp\left\{\mathrm{i}\omega_1 t\left(\hat{b}^\dagger\hat{b} + \frac{1}{2}\hat{\sigma}_z\right)\right\} \tag{6.3}$$

we move into the frame rotating with ω_1 and the transformed Hamiltonian becomes

$$\begin{aligned}\frac{H_1}{\hbar} =& \frac{\epsilon}{2}\hat{\sigma}_z + \omega\hat{b}^\dagger\hat{b} + g\left(\hat{\sigma}_+ e^{\mathrm{i}\omega_1 t} + \hat{\sigma}_-^{-\mathrm{i}\omega_1 t}\right)\left(e^{\mathrm{i}\omega_1 t}\hat{b}^\dagger + e^{-\mathrm{i}\omega_1 t}\hat{b}\right) \\ &+ \eta_1\left(\hat{\sigma}_+ e^{\mathrm{i}\omega_1 t} + \hat{\sigma}_- e^{-\mathrm{i}\omega_1 t}\right)\cos\omega_1 t + \eta_2\left(\hat{\sigma}_+ e^{\mathrm{i}\omega_1 t} + \hat{\sigma}_- e^{-\mathrm{i}\omega_1 t}\right)\cos\omega_2 t \\ &- \omega_1\left(\hat{b}^\dagger\hat{b} + \frac{1}{2}\hat{\sigma}_z\right) \qquad (6.4)\\ =& \frac{1}{2}(\epsilon - \omega_1)\hat{\sigma}_z + (\omega - \omega_1)\hat{b}^\dagger\hat{b} + g\left(\hat{\sigma}_-\hat{b}^\dagger + \hat{\sigma}_+\hat{b}\right) + \frac{\eta_1}{2}(\hat{\sigma}_+ + \hat{\sigma}_-) \\ &+ \frac{\eta_2}{2}\left(\hat{\sigma}_+ e^{\mathrm{i}(\omega_1-\omega_2)t} + \hat{\sigma}_- e^{-\mathrm{i}(\omega_1-\omega_2)t}\right) + g\left(\hat{\sigma}_+ e^{2\mathrm{i}\omega_1 t}\hat{b}^\dagger + \hat{\sigma}_- e^{-2\mathrm{i}\omega_1 t}\hat{b}\right) \\ &+ \frac{\eta_1}{2}\left(\hat{\sigma}_+ e^{2\mathrm{i}\omega_1 t} + \hat{\sigma}_- e^{-2\mathrm{i}\omega_1 t}\right) + \frac{\eta_2}{2}\left(\hat{\sigma}_+ e^{\mathrm{i}(\omega_1+\omega_2)t} + \hat{\sigma}_- e^{\mathrm{i}(-\omega_1-\omega_2)t}\right) \qquad (6.5)\\ \approx& (\epsilon - \omega_1)\frac{\hat{\sigma}_z}{2} + (\omega - \omega_1)\hat{b}^\dagger\hat{b} + g\left(\hat{\sigma}_-\hat{b}^\dagger + \hat{\sigma}_+\hat{b}\right) \\ &+ \frac{\eta_1}{2}\hat{\sigma}_x + \frac{\eta_2}{2}\left(\hat{\sigma}_+ e^{\mathrm{i}(\omega_1-\omega_2)t} + \hat{\sigma}_- e^{-\mathrm{i}(\omega_1-\omega_2)t}\right). \qquad (6.6)\end{aligned}$$

Terms of the form $e^X Y e^{-X}$ are calculated using the power series expansion of the exponential function, also called Hadamard lemma. For $X = \mathrm{i}\omega_1 t\hat{b}^\dagger\hat{b}$, $Y = \hat{b}^\dagger$,

$$e^{\mathrm{i}\omega_1 t\hat{b}^\dagger\hat{b}}\hat{b}^\dagger e^{-\mathrm{i}\omega_1 t\hat{b}^\dagger\hat{b}} = e^{\mathrm{i}\omega_1 t}\hat{b}^\dagger \qquad (6.7)$$

$$e^{\mathrm{i}\omega_1 t\hat{b}^\dagger\hat{b}}\hat{b}e^{-\mathrm{i}\omega_1 t\hat{b}^\dagger\hat{b}} = e^{-\mathrm{i}\omega_1 t}\hat{b} \qquad (6.8)$$

since

$$\sum_{m=0}^{\infty}\frac{1}{m!}\left[\mathrm{i}\omega_1 t\hat{b}^\dagger\hat{b}, \hat{b}^\dagger\right]_m = \hat{b}^\dagger\sum_{m=0}^{\infty}\frac{1}{m!}(\mathrm{i}\omega_1 t)^m = \hat{b}^\dagger e^{\mathrm{i}\omega_1 t}. \qquad (6.9)$$

$e^{\frac{\mathrm{i}}{2}\omega_1 t\hat{\sigma}_z}$ and $\hat{\sigma}_z$ clearly commute, while

$$e^{\frac{\mathrm{i}}{2}\omega_1 t\hat{\sigma}_z}\hat{\sigma}_x e^{-\frac{\mathrm{i}}{2}\omega_1 t\hat{\sigma}_z} = \hat{\sigma}_+ e^{\mathrm{i}\omega_1 t} + \hat{\sigma}_- e^{-\mathrm{i}\omega_1 t} \qquad (6.10)$$

using $e^{\mathrm{diag}(a,b)} = \mathrm{diag}(e^a, e^b)$. Terms rotating at $e^{\pm 2\mathrm{i}\omega_1 t}$ were omitted in Eq. (6.6) in the spirit of a RWA, valid for $\eta_1/2\omega_1 \ll 1$. This is a good approximation as η_1 is bound in the experiment by the qubit anharmonicity.

The η_1-term is now the significant term and we move into its interaction picture via the transformation

$$e^{\mathrm{i}\frac{\eta_1}{2}\hat{\sigma}_x t}\left[\frac{\hat{H}_1}{\hbar} - \frac{\eta_1}{2}\hat{\sigma}_x\right]e^{-\mathrm{i}\frac{\eta_1}{2}\hat{\sigma}_x t}. \qquad (6.11)$$

Satisfying the requirement $\omega_1 - \omega_2 = \eta_1$ and performing a time averaging in the spirit of a RWA yields the effective Hamiltonian in the ω_1 frame

$$\frac{\hat{H}_{\mathrm{eff}}}{\hbar} = \frac{\eta_2}{2}\frac{\hat{\sigma}_z}{2} + \omega_{\mathrm{eff}}\hat{b}^\dagger\hat{b} + \frac{g}{2}\hat{\sigma}_x\left(\hat{b}^\dagger + \hat{b}\right). \qquad (6.12)$$

We define the effective bosonic mode energy $\omega_{\text{eff}} \equiv \omega - \omega_1$, which is the parameter governing the system dynamics. Noting that $\eta_1 \gg \eta_2$, which is a necessary condition for the above approximation to hold, the effective qubit frequency η_2 and effective bosonic mode frequency ω_{eff} can be chosen as experimental parameters in the simulation. In Eq. (6.12), the complete coupling term of the quantum Rabi Hamiltonian is recovered, valid in the USC regime and beyond, while the geometric coupling strength is only modified by a factor of two, resulting in $g_{\text{eff}} = g/2$. It is therewith feasible to tune the system into a regime where the coupling strength is comparable to the energies of the subsystems. This is achieved by leaving the geometric coupling strength essentially unchanged in the synthesized Hamiltonian, while slowing down the system dynamics by effectively decreasing the mode frequencies to $\lesssim 8\,\text{MHz}$.

Table 6.1: Designation of coupling regimes in the quantum Rabi model The coupling strength g must exceed the decay rates Γ_1, κ of the coupled subsystems for the strong coupling condition from cavity QED to be fulfilled, see Fig. 6.3. For USC and deep strong coupling, the figure of merit is the coupling strength relative to the bosonic mode frequency, g/ω.

coupling regime	condition
strong coupling	$g > \Gamma_1, \kappa$
USC	$0.1 < g/\omega < 1$
deep strong	$g/\omega > 1$

Thermal excitations of these effective transitions can be neglected since they couple to the thermal bath excitation frequency of the cryostat, corresponding to $\sim 1\,\text{GHz}$, via their laboratory frame equivalent frequency of $\omega/2\pi \sim 6\,\text{GHz}$. We want to point out that the coupling regime is defined by $g_{\text{eff}}/\omega_{\text{eff}}$, rather than by the Rabi frequency η_1, which does not enter the synthesized Hamiltonian. See Tab. 6.1 for an overview on the designation of coupling regimes.

While the simulation scheme requires $|\epsilon - \omega_1| \ll \eta_1$, the qubit frequency ϵ does not enter the effective Hamiltonian. The time evolution of the qubit measured in the laboratory frame is subject to fast oscillations corresponding to the Rabi frequency η_1. Accordingly, the qubit dynamics in the engineered quantum Rabi Hamiltonian Eq. (6.12), valid in the ω_1 frame, can be inferred from the envelope of the evolution in the laboratory frame. The derivation of Eq. (6.12) can be found in Ref. [Bal+12] and is detailed in Supplementary Note 1 of Ref. [Bra+17]. A similar drive scheme based on a Rabi tone was previously used in experiment to synthesize an effective Hamiltonian with a rotated qubit basis [Voo+16]. For the qubit and the bosonic mode degenerate in the laboratory frame, a distinct collapse-revival signature appears in the dynamics of the quantum Rabi model under USC conditions, which is observed and investigated in this experiment.

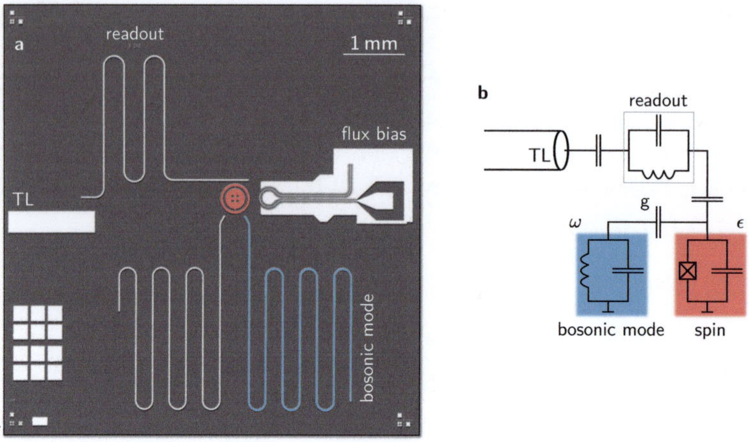

Figure 6.1: Quantum simulation device (a) Optical micrograph with the atomic spin represented by a concentric transmon qubit, highlighted in red and the $\lambda/2$ microstrip resonator (blue) constituting the bosonic oscillator mode. The readout resonator couples to the qubit capacitively and its frequency is measured with an open transmission line (TL) via the reflection signal of an applied microwave tone or pulse. The second resonator visible on chip is not used in the current experiment and is detuned in frequency from the relevant bosonic mode by $\sim 0.5\,\mathrm{GHz}$. (b) Effective circuit diagram of the device.

6.3 Device

The physical implementation of the quantum simulator is based on a superconducting circuit embedded in a typical circuit QED setup [Bla+04; Wal+04], see Fig. 6.1. The atomic spin of the quantum Rabi model is mapped to a concentric transmon qubit [Koc+07; Bra+16]. It is operated at a ratio of Josephson energy to charging energy $E_J/E_C = 50$ and an anharmonicity $\alpha/h = \epsilon_{12}/2\pi - \epsilon_{01}/2\pi = -0.36\,\mathrm{GHz} \sim -E_C/h = -0.31\,\mathrm{GHz}$, and tuned on resonance with the bosonic mode at 5.948 GHz. ϵ_{ij} denote the transition frequencies between transmon levels i, j. The energy relaxation rate Γ_1 of the qubit at the operation point is measured to be $\Gamma_1 = 1/T_1 = 0.2 \times 10^6\,\mathrm{s}^{-1}$. An on-chip flux bias line allows for a fast tuning of the qubit transition frequency, see Ch. 5. The bosonic mode of the model is represented by a harmonic $\lambda/2$ resonator with an inverse lifetime $\kappa \sim 3.9 \times 10^6\,\mathrm{s}^{-1}$ that is limited by internal loss.

6.3.1 Fabrication

Sample fabrication was carried out in one single electron beam lithography step. The Josephson junctions are formed by shadow angle evaporation and the Dolan bridge technique, as detailed in Ch. 4. The area of the Josephson junctions is designed to be $100\,\text{nm} \times 220\,\text{nm}$, resulting in a critical current $I_c = 45\,\text{nA}$ for a single junction. We applied an Al metallization on the backside of the double-side polished intrinsic Si substrate as a ground reference for the microstrip elements on chip.

6.3.2 Measurement technique

Qubit excitation and Rabi driving are performed by heterodyne mixing with two individual IQ frequencies and amplitudes, respectively. In the experiment with only the dominant Rabi drive applied, we use one single microwave source and one IQ mixer according to the setup shown in Fig. 4.2 in Ch. 4. This allows for the required phase control on the phase φ_1 of the dominant Rabi drive with respect to the qubit Bloch sphere coordinate system fixed by the first excitation pulse. In particular, we fix the idling time between initial excitation pulse and the onset of the Rabi drive, such that the acquired phase during that time is constant. The Rabi drive is therefore applied with a constant relative phase φ_1 with respect to the phase used for the excitation pulse. We chose an LO frequency located 20 MHz or 65 MHz above the qubit control frequency, located at about $\omega/2\pi + 95\,\text{MHz}$.

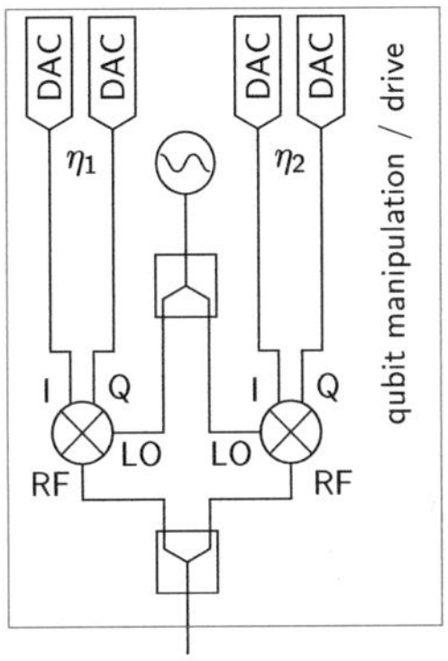

Figure 6.2: Qubit manipulation and drive microwave setup Modification of the qubit manipulation setup, see the grey box in Fig. 4.2. Two drive pulses with amplitudes η_1, η_2, respectively, are generated in separate IQ mixers, sharing a common local oscillator (LO) input, and combined subsequently.

In the experiment with the second Rabi drive added, both drive tones are up-converted in two separate IQ mixers while sharing a common local oscillator input to preserve their relative phase relation, see Fig. 6.2(a). Qubit excitation and drive pulses are applied via the same transmission line used for readout.

The bosonic mode resonator is located far away from the transmission line in space, which reduces unwanted parasitic driving. It is strongly detuned from the readout resonator located at $\omega_{\mathrm{r}}/2\pi = 8.86\,\mathrm{GHz}$.

6.4 Sample characterization

The quantum state collapse followed by a quantum revival is the most striking signature of the ultra-strong coupling regime of the quantum Rabi model. It emerges for qubit and bosonic mode degenerate in the laboratory frame. We calibrate this resonance condition by minimizing the periodic swap rate of a single excitation between qubit and bosonic mode for the simple Jaynes-Cummings model in the absence of additional Rabi drives. Figures 6.3(a), (b) show the measured vacuum Rabi oscillations in the resonant case (a) and dependent on the qubit transition frequency (b). For qubit preparation and readout we detune the qubit by 95 MHz up in frequency and therewith out of resonance. This corresponds to switching off the resonant interaction with the bosonic mode. Experimental details on flux pulse generation are described in Sec. 4.3. Vacuum Rabi oscillations can be observed during the interaction time Δt and yield a coupling strength $g/2\pi = 4.3\,\mathrm{MHz}$.

Table 6.2: Summary of device and simulation parameters Errors are extracted from the fit of the vacuum Rabi oscillations in Fig. 6.3.

$\omega/2\pi$	$E_{\mathrm{J}}/E_{\mathrm{C}}$	α/h	$\Gamma_1 = 1/T_1$	κ
5.948 GHz	50	-0.36 GHz	$(0.2 \pm 0.12) \times 10^6\,\mathrm{s}^{-1}$	$(3.9 \pm 0.14) \times 10^6\,\mathrm{s}^{-1}$

$\omega_{\mathrm{r}}/2\pi$	$\omega_{\mathrm{eff}}/2\pi$	$\eta_1/2\pi$	$\eta_2/2\pi$
8.86 GHz	< 8 MHz	~ 50 MHz	~ 3 MHz

This is in good agreement with the spectroscopically obtained result, see Fig. 6.3. A two-tone spectroscopy measurement yields a minimum line separation of $2g/2\pi = 8.0\,\mathrm{MHz}$, corresponding to a coupling strength of $g/2\pi = 4.0\,\mathrm{MHz}$. The dispersive readout resonator shift is measured with a continuous microwave probe tone and an additional microwave drive tone is applied through the same transmission line to excite the qubit transition.

The relevant experimental device and simulation parameters are summarized in Tab. 6.2.

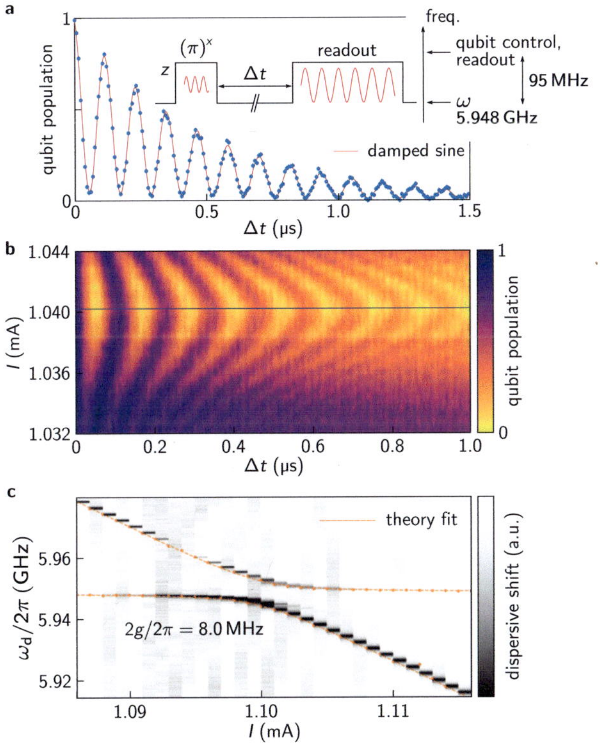

Figure 6.3: Calibration of the coupling strength g between qubit and bosonic mode (a) The qubit is initially dc-biased on resonance with the bosonic mode, while it is detuned for state preparation and readout. The solid black line in the inset depicts the fast flux pulses applied to the flux bias line and indicates the qubit frequency on the given axis. After preparing the qubit in $|e\rangle$, qubit and bosonic mode are on resonance during an interaction time Δt. A frequency fit (red) of the vacuum Rabi oscillations yields $2g/2\pi = 8.5\,\mathrm{MHz}$. With the decay rate $\Gamma = (2.08 \pm 0.03) \times 10^6\,\mathrm{s}^{-1}$ of the envelope and the qubit decay rate $1/T_1 = (0.2 \pm 0.12) \times 10^6\,\mathrm{s}^{-1}$ we extract the bosonic mode decay rate $\kappa = (3.9 \pm 0.13) \times 10^6\,\mathrm{s}^{-1}$. Error bars denote a statistical standard deviation. (b) Typical calibration measurement for the resonance condition (blue line) between qubit and bosonic mode. By varying the dc bias current I, we can tune out of resonance and observe the expected decrease in excitation swap efficiency and an increase in the vacuum Rabi frequency. The qubit population is given in colours and we applied a numerical interpolation of data points. (c) Spectroscopic measurement of the avoided crossing between qubit and bosonic mode. The qubit transition frequency is tuned by a dc current applied to the flux bias line. The dispersive shift of the readout resonator is proportional to the excitation number of the qubit and is depicted in colours. A fit yields a minimum line separation of $2g/2\pi = 8.0\,\mathrm{MHz}$.

6.5 Master equation simulations

Master equation simulations of the quantum Rabi model are based on the master equation solver provided by the QuTiP package [JNN12; JNN13] for Python. Here, the time evolution of a given initial state or density matrix is calculated by solving the von Neumann equation associated with the given system Hamiltonian in the absence of dissipation. Dissipation is taken into account by solving the Lindblad master equation.

The decay terms in the underlying Liouvillian are based on the experimentally found parameters, summarized in Tab. 6.2. In addition we use a qubit dephasing time of $T_2^* \sim 0.5\,\mu\mathrm{s}$. The Hilbert space of the bosonic mode is truncated at a photon number of 25 to 35, while higher Fock states were found to be insignificant. The transmon qubit is treated as a three-level system with the experimentally found anharmonicity $\alpha/h = -0.36\,\mathrm{GHz}$. To account for the $|\mathrm{f}\rangle$ level in the transmon, we use a transversal qubit coupling operator

$$\sum_{ij} g_{ij} \left|i\right\rangle \left\langle j\right| . \tag{6.13}$$

The coupling matrix elements g_{ij} are found by evaluating the matrix elements of the Cooper pair number operator $\hat{N}$ in the charge basis [Koc+07; Bra+15], according to

$$g_{ij} \propto |\langle i|\hat{N}|j\rangle|. \tag{6.14}$$

In analogy, we can define a general annihilation operator $\hat{a}_\mathrm{t}$ that reflects the transmon anharmonicity in the simulation according to

$$\hat{a} = \sum_{i<j} \frac{g_{ij}}{g_{01}} \left|i\right\rangle \left\langle j\right| . \tag{6.15}$$

Master equation simulations of the effective USC quantum Rabi model in the rotating frame are performed in the two-level approximation.

6.6 Verification of the simulation scheme

Since a rotating frame is not an inertial system, the laws of physics may be drastically altered when describing a physical system in a rotating frame. In the framework of analog quantum simulation, this allows us to access intriguing effective parameter regimes that are hard or impossible to tread in the laboratory frame. For the simulation scheme applied here, it is therefore not a priori clear that the bosonic mode population evolution of the quantum Rabi model is well reflected by the laboratory frame simulation. To verify the simulation scheme for the ideal system we compare the bosonic mode population in the driven laboratory frame, and the expected evolution of the ideal effective quantum Rabi model via master equation simulations. Here, we neglect dissipation and any parasitic drive terms in the Hamiltonian. Figure 6.4(a) demonstrates good agreement when comparing the time evolutions of the ideal (solid line) and the constructed (dashed line) Hamiltonians for various choices of the bosonic mode frequency ω_{eff} in the effective frame.

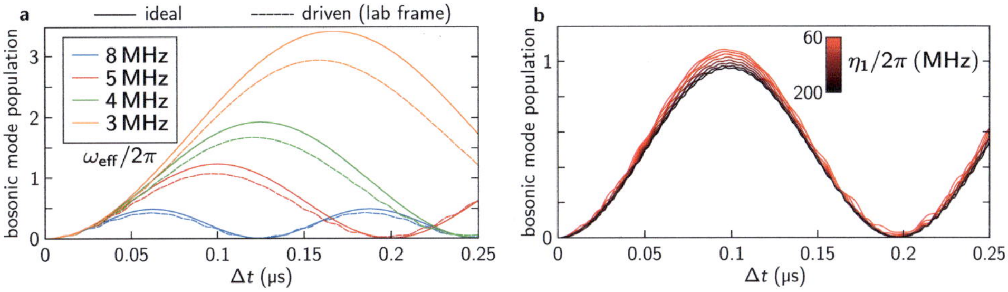

Figure 6.4: Verification of the simulation scheme and effect of the parasitic drive (a) Time evolution of the bosonic mode population for a simulation sequence with $\eta_2 = 0$. We compare its population in the ideal quantum Rabi Hamiltonian (solid line) with the population of the bosonic mode in the laboratory frame with the drive applied (dashed line). Despite the fact that an infinite energy reservoir is supplied by the drive, the population follows the expected one rather well. This remains also true for varying the drive amplitude η_1. Different colours correspond to a varying ω_{eff}. (b) Bosonic mode population for $\omega_{\text{eff}}/2\pi = 5\,\text{MHz}$ and a varying drive amplitude η_1. The evolution and maximum population is confirmed to be independent of η_1 in first order. Master equation simulations here are performed without taking into account dissipation and neglect parasitic driving of the bosonic mode.

For $\omega_{\text{eff}}/2\pi = 5\,\text{MHz}$ we demonstrate that the photon population in the bosonic mode is independent of the applied drive amplitude η_1, despite of it forming a large energy reservoir that is provided to the circuit. Simulations for various η_1 in the driven laboratory frame are depicted in Fig. 6.4(b).

6.7 Quantum state collapse and revival

The collapse-revival signature of the quantum Rabi model in USC conditions manifests most clearly for a vanishing qubit term. We therefore initially perform a simulation with $\eta_2 = 0$, yielding an effective Hamiltonian in the qubit frame

$$\frac{\hat{H}}{\hbar} = \omega_{\text{eff}}\hat{b}^{\dagger}\hat{b} + \frac{g}{2}\hat{\sigma}_x\left(\hat{b}^{\dagger} + \hat{b}\right). \tag{6.16}$$

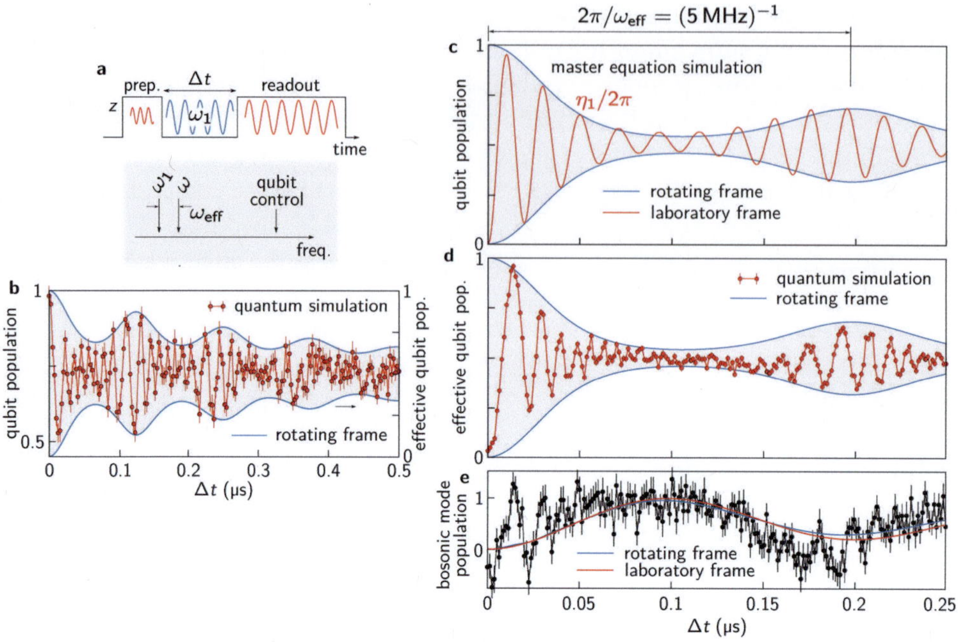

Figure 6.5: Quantum state collapse and revival with only the dominant Rabi drive applied (a) Schematic pulse sequence and overview on the relative frequencies used in the experiment. (b) Quantum simulation of the periodic recurrence of quantum state revivals for $\omega_{\text{eff}}/2\pi = 8\,\text{MHz}$. The blue line corresponds to a master equation simulation of the qubit evolution in the rotating frame. (c), (d) Master equation and quantum simulation of the qubit time evolution for initial qubit states $|g\rangle$, $|e\rangle$ and $\omega_{\text{eff}}/2\pi = 5\,\text{MHz}$, corresponding to $g_{\text{eff}}/\omega_{\text{eff}} \sim 0.5$. The red line shows the qubit population evolution of the driven system in the laboratory frame, Eq. (6.2), while the blue lines follow the qubit evolution in the synthesized Hamiltonian Eq. (6.12), likewise extracted from a master equation simulation. The deviation between the envelope of the laboratory frame data and the rotating frame data in (c) reflects the approximations of the simulation scheme. Experimental data shows the difference between two measurements for the qubit prepared in $|g\rangle$, $|e\rangle$, respectively, in order to isolate the qubit signal. (e) Measured population evolution of the bosonic mode, extracted from the sum of the two successive measurements and fitted to simulated data. Error bars throughout the figure denote a statistical standard deviation of the mean and are propagated according to Gauss.

Figure 6.5(a) shows the applied measurement sequence which is based on the one in Fig. 6.3(a) but extended by a drive tone of amplitude η_1. The bosonic mode is initially in the vacuum state and the qubit is prepared in one of its basis states $|g\rangle$, $|e\rangle$, which are thermally impure. Qubit and bosonic mode are on resonance during the simulation time Δt. The effective bosonic mode frequency ω_{eff} in the rotating frame is set by the detuning of the Rabi drive at a frequency ω_1 from the common resonance frequency ω. Measured data for $\omega_{\text{eff}}/2\pi = 8\,\text{MHz}$ is displayed in Fig. 6.5(b), corresponding to $g_{\text{eff}}/\omega_{\text{eff}} \sim 0.3$. Data points show the experimentally simulated time evolution of the qubit prepared in $|e\rangle$. A fast quantum state collapse followed by periodically returning quantum revivals can be observed. The ground state of the qubit subspace in the driven system as well as in the synthesized Hamiltonian, Eq. (6.16), is in the equatorial plane of the qubit Bloch sphere and is occupied after a time $\Delta t \gg T_1,\ 1/\kappa$. It is diagonal in the $|\pm\rangle$ basis, with $|\pm\rangle = 1/\sqrt{2}(|e\rangle \pm |g\rangle)$.

The revival dynamics can be understood with an intuitive picture in the laboratory frame. The eigenenergies in the $|\pm\rangle$ subspaces take the form of displaced vacuum

$$\omega_{\text{eff}} \left(\hat{b}^{\dagger} \pm \frac{g}{2\omega_{\text{eff}}}\right) \left(\hat{b} \pm \frac{g}{2\omega_{\text{eff}}}\right) + \text{const.}, \tag{6.17}$$

which is a coherent state that is not diagonal in the Fock basis. The prepared initial state in the experiment is therefore not an eigenstate in the effective basis of the driven system. Consequently, many terms that correspond to the relevant Fock states n of the bosonic mode participate in the dynamics with phase factors $\exp\{in\omega_{\text{eff}}t\}$, $n \in \mathbb{N}^{+}$. While contributing terms get out of phase during the state collapse, they rephase after an idling period of $2\pi/\omega_{\text{eff}}$ to form the quantum revival. The underlying physics of this phenomenon is fundamentally different from the origin of state revivals that were proposed for the Jaynes-Cummings model [ENS80]. Here, the preparation of the bosonic mode in a large coherent state with $\alpha \gtrsim 10$ is required and non-periodic revivals are expected at times $\propto 1/g_{\text{eff}}$ rather than $\propto 1/\omega_{\text{eff}}$ [HR06], as demonstrated in Fig. 6.6.

The blue line in Fig. 6.5(b) corresponds to a master equation simulation of the qubit dynamics in the effective frame, Eq. (6.16).

Figures 6.5(c),(d) show a master equation simulation and the quantum simulation for $\omega_{\text{eff}}/2\pi = 5\,\text{MHz}$, respectively, with the qubit prepared in one of its eigenstates $|g\rangle$, $|e\rangle$. The population of the bosonic mode takes a maximum during the idling period and adopts its initial population at $2\pi/\omega_{\text{eff}}$ in the absence of dissipation, see Fig. 6.5(e). The fast oscillations in Fig. 6.5(c), (d) correspond to the Rabi frequency $\eta_1/2\pi \sim 50\,\text{MHz}$. This value is chosen such that the requirement $\eta_1/\omega_{\text{eff}} \gg 1$ is fulfilled while staying well below the transmon anharmonicity, avoiding higher level populations.

Deviations in the laboratory frame simulation traces are due to an uncertainty in the Rabi frequency that is extracted from Fourier transformation of measured data, see Fig 6.7(c). The broadening in frequency space is mainly caused by the beating in experimental data, which is an experimental artefact, potentially originating from pulse imperfections. The relevant dynamics of the USC quantum Rabi Hamiltonian corresponds to the envelope of measured data.

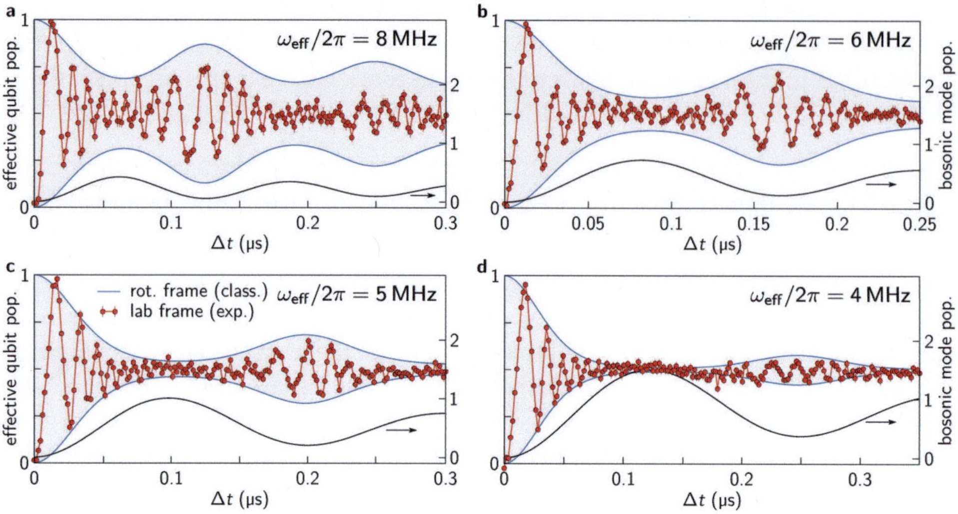

Figure 6.6: Quantum collapse and revival signatures for various ω_{eff} The qubit is prepared in its groundstate $|g\rangle$ and the bosonic mode is initially in the vacuum state $|0\rangle$. The first revival appears at $2\pi/\omega_{\text{eff}}$, respectively, and the different plots correspond to a varying ω_{eff}. The blue line shows a master equation simulation of the ideal effective Hamiltonian in the rotating frame, while the red data points are the qubit population evolution in the effective qubit frame. The black line shows the bosonic mode population in the rotating frame according to a master equation simulation, which increases with decreasing ω_{eff}. Note that the scale on the horizontal axis is not equal for each plot. The depicted qubit signal is extracted from measured data with the protocol described in Sec.6.7.1, based on the master equation simulation (black) in the absence of a parasitic drive of the bosonic mode.

Since the laboratory frame dissipation is enhanced for a larger ratio of photon population in the bosonic mode, the accessible coupling regime is bound by the limited coherence of the bosonic mode, in particular. This is demonstrated by a dependence of the coherence envelope of the quantum revivals on the ratio g/ω_{eff}, see Fig. 6.6. The dependence reflects that the excitation number is no longer a conserved quantity in the quantum Rabi model at USC conditions. Figure 6.6 further demonstrates that the maximum position of the first quantum revival appears at to $2\pi/\omega_{\text{eff}}$, which can be well controlled by choosing ω_{eff} in the quantum simulation.

We find a better agreement of experimental data with master equation simulations for a slightly increased geometric coupling $g/2\pi \sim 5.5\,\text{MHz}$ as compared to the measured value in Fig. 6.3 of $g/2\pi = 4.3\,\text{MHz}$. This may be explained by an excess population of the bosonic mode due to a parasitic coupling of the Rabi drives, see Sec. 6.7.2. This can effectively increase the coupling strength between qubit and bosonic mode. The effect may be additionally enhanced by a population of higher transmon levels.

Table 6.3: Geometric coupling $g/2\pi$ between qubit and bosonic mode. Experimental values are obtained from the spectroscopic measurement and the vacuum Rabi experiment in Fig. 6.3. The value used in master equation simulations differs slightly.

spectroscopy	vacuum Rabi	master eq. simulation
4.0 MHz	4.3 MHz	5.5 MHz

We checked that master equation simulations with increased decay and decoherence rates, accounting for a changed environmental spectral density in the rotating frame, cannot reproduce the correct dynamics. Simulations performed with $g/2\pi =$ 4.3 MHz show a $\sim$ 30% decrease in the maximum population of the bosonic mode with no qualitative consequences. With the measured geometric coupling we approach an USC regime with $g_{\text{eff}}/\omega_{\text{eff}} \sim 0.6$ for the experiment depicted in Fig. 6.6(d). Based on the effectively increased value for g_{eff}, we reach a relative coupling ratio of 0.7. Measured values of g_{eff} and the effective value used in the presented master equation simulations are summarized in Tab. 6.3.

6.7.1 Protocol for extracting the qubit population

In the simulation experiments presented in Fig. 6.5, we note a modulated low-frequency bulge in the recorded dispersive readout resonator shift that does not agree with the expected qubit population evolution, see Fig. 6.7(a). By comparing with the master equation simulation, we can recognize the population evolution of the bosonic mode which reflects the governing fundamental frequency ω_{eff} of the effective Hamiltonian. By simulating the full circuit Hamiltonian including qubit, bosonic mode and readout resonator, we find that the effect is induced by an additional photon exchange coupling f between the bosonic mode and the readout resonator. The coupling is facilitated by the electric fields of the resonators and is potentially mediated by the qubit. By inheriting non-linearity from the qubit, the bosonic mode can induce a cross-Kerr like photon number dependent shift $\propto f^2$ on the harmonic readout resonator.

The complete Hamiltonian including the readout resonator of resonance frequency ω_{r} with creation (annihilation) operator $\hat{a}^\dagger$ ($\hat{a}$) and with the RWA applied takes the form

$$\begin{aligned}\frac{\hat{H}}{\hbar} =& \frac{\epsilon}{2}\hat{\sigma}_z + \omega\hat{b}^\dagger\hat{b} + \omega_{\text{r}}\hat{a}^\dagger\hat{a} + g\left(\hat{\sigma}_-\hat{b}^\dagger + \hat{\sigma}_+\hat{b}\right) + g_{\text{r}}\left(\hat{\sigma}_-\hat{a}^\dagger + \hat{\sigma}_+\hat{a}\right) \\ &+ f\left(\hat{a}\hat{b}^\dagger + \hat{a}^\dagger\hat{b}\right) + \hat{\sigma}_x\eta_1\cos(\omega_1 t + \varphi_1) + \hat{\sigma}_x\eta_2\cos(\omega_2 t + \varphi_2).\end{aligned} \tag{6.18}$$

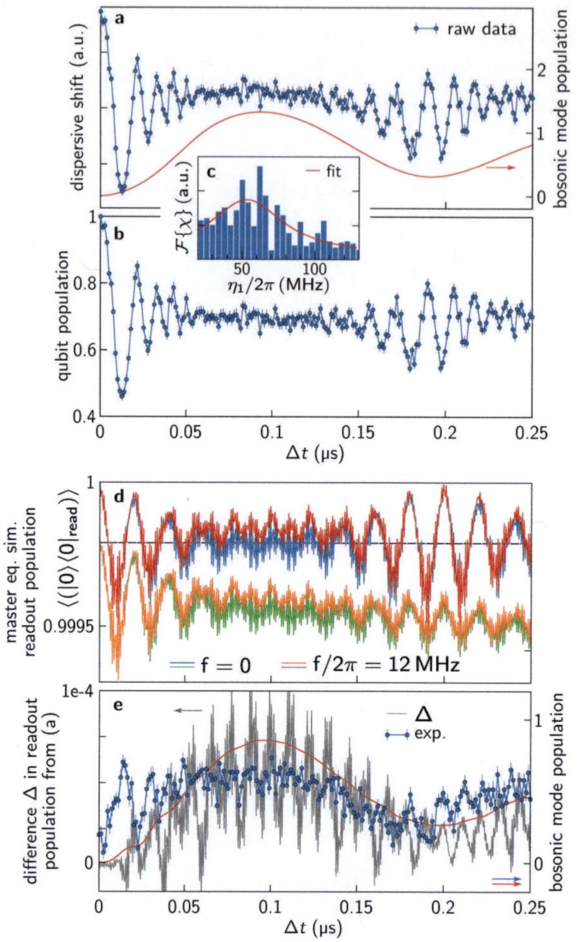

Figure 6.7: Verification of the photon exchange coupling between the bosonic mode and the readout resonator (a) Dispersive shift of the readout resonator including both the shifts induced by qubit and bosonic mode. The red line depicts the bosonic mode population according to a master equation simulation. (b) Extracted qubit signal in the original qubit basis after subtracting the additional shift induced by the bosonic mode. The analysis is based on the expectation from the master equation simulation as depicted in (a). The qubit was prepared in $|e\rangle$. From Fourier transformation (c) we obtain $\eta_1/2\pi \sim 52\,\text{MHz}$. (d) Master equation simulation of the vacuum projection $|0\rangle\langle 0|$ of the readout resonator state for vanishing photon exchange coupling $f = 0$ (blue) and for $f/2\pi = 12\,\text{MHz}$ (red), disregarding dissipation. The blue dashed line denotes the mean value of the $f = 0$ simulation as a guide to the eye. Orange and green traces show the same master equation simulations in the presence of dissipation and are shifted for better visibility. The additional bulge in the presence of the photon exchange coupling $f > 0$ is apparent. (e) The difference (grey) between both master equation simulations from (d) including dissipation follows the experimentally measured bosonic mode population (blue). Both data sets are fitted to the time evolution of the bosonic mode population (red), as obtained from the same master equation simulation.

$g_r/2\pi \sim 55$ MHz denotes the coupling strength between qubit and readout resonator. The conjecture is verified by comparing master equation simulations with the photon exchange coupling f switched on and off, respectively, see 6.7(d). As visible in 6.7(e), the difference of both master equation simulations (grey) follows the evolution of the bosonic mode population (red) in the rotating frame as obtained from the same simulation. The experimentally extracted bosonic mode population (blue) likewise agrees with the trend from the master equation simulation. The photon exchange coupling assumes the form of a cross-Kerr interaction after diagonalization in the subspace spanned by the bosonic mode and the readout resonator. Comparison with measured data suggests $f \sim$ MHz.

In Fig. 6.5(d) we made use of the topological symmetry of simulations with initial qubit states $|g\rangle$, $|e\rangle$. In order to cancel out the additional dispersive shift induced by the bosonic mode, we subtract two successive measurements with the qubit prepared in its eigenstates $|g\rangle$, $|e\rangle$, respectively. The population evolution of the bosonic mode, depicted in Fig. 6.5(e), is obtained by summing two successive measurements with the qubit prepared in $|g\rangle$, $|e\rangle$, respectively. The extracted bosonic mode population is fitted to the expectation according to master equation simulations in the laboratory frame and in the effective frame. Since the maximum population is around unity while the qubit is in the equatorial state, the non-conservation of the total excitation number is apparent.

This measurement protocol is based on the symmetry of the qubit signal for preparing eigenstates, while the bosonic mode induced shift is always repulsive and does not change its sign. The photon exchange coupling f therefore provides indirect access to the population of the bosonic mode without a dedicated readout device available. Specifically monitoring the population of the bosonic mode and performing a Wigner tomography would highlight another hallmark signature of the USC regime, namely the efficient generation of non-classical cavity states [Lan+17]. The protocol for generating cat states in the bosonic mode with the presented scheme is elaborated in Sec. 6.9.

In the absence of such a symmetry, the qubit population can be retrieved from measured raw data based on the master equation simulation of the bosonic mode population. In this procedure, the dispersive shift $\propto f^2$ remains as the only free fit parameter. Figure 6.7(b) shows the extracted qubit signal after subtracting the additional shift induced by the bosonic mode. The mentioned symmetry is absent when the relative phases of the Rabi drives are relevant or when we apply the second Rabi drive.

6.7.2 Parasitic driving of the bosonic mode

In the experiment we face a parasitic coupling of the Rabi drive tones to the bosonic mode that is degenerate with the qubit and spatially close by in the circuit. This leads to an excess population of the bosonic mode, however without disturbing the functional evolution of its population.

In our circuit, the effect is accounted for by an additional drive term

$$\eta_{\mathrm{r}}\left(\hat{b}^{\dagger}+\hat{b}\right)\cos(\omega_1 t) \tag{6.19}$$

entering the laboratory frame Hamiltonian, Eq. (6.2). η_{r} denotes the effective amplitude of the parasitic drive and we take into account the effect of only the dominant Rabi drive frequency ω_1. Under the unitary transformation $\hat{U}$ from Eq. (6.3), Eq. (6.19) becomes

$$\frac{\eta_{\mathrm{r}}}{2}\left(\hat{b}^{\dagger}+\hat{b}\right)+\frac{\eta_{\mathrm{r}}}{2}\left(e^{2\mathrm{i}\omega_1 t}\hat{b}^{\dagger}+e^{-2\mathrm{i}\omega_1 t}\hat{b}\right). \tag{6.20}$$

Terms rotating with $e^{\pm 2\mathrm{i}\omega_1 t}$ are omitted, resulting in a time independent drive term that is added to the Hamiltonian (6.12) in the rotating frame. The effective Hamiltonian including the parasitic drive term reads

$$\frac{\hat{H}_{\mathrm{eff,p}}}{\hbar}=\frac{\eta_2}{2}\frac{\hat{\sigma}_z}{2}+\omega_{\mathrm{eff}}\hat{b}^{\dagger}\hat{b}+\frac{g}{2}\sigma_x\left(\hat{b}^{\dagger}+\hat{b}\right)+\frac{\eta_{\mathrm{r}}}{2}\left(\hat{b}^{\dagger}+\hat{b}\right). \tag{6.21}$$

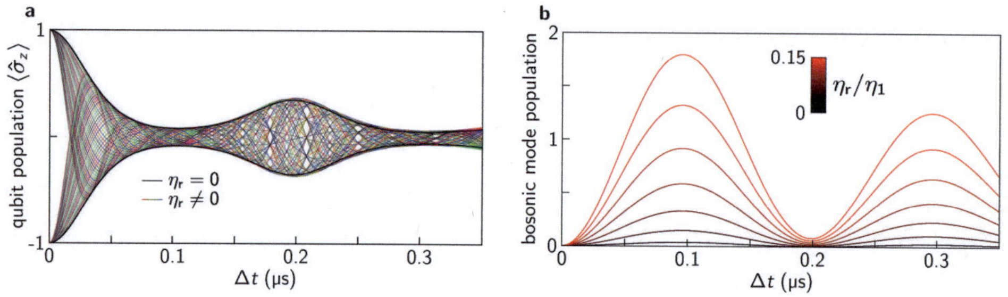

Figure 6.8: Effect of parasitic driving of the bosonic mode (a) Master equation simulation of the qubit population for the effective Hamiltonian, Eq. (6.24), without the qubit tunneling term $\eta_{\mathrm{r}}=0$ (black) in comparison with simulations for $\eta_{\mathrm{r}}\neq 0$, plotted in colours. One can see that the additional qubit tunneling term $\propto \hat{\sigma}_x$ introduces a rotation while complying with the envelope of the ideal Hamiltonian, see Eq. (6.12). We demonstrate this by plotting the evolution for various values of η_{r} and $\omega_{\mathrm{eff}}/2\pi = 5\,\mathrm{MHz}$. (b) Bosonic mode population based on a master equation simulation for a harmonic oscillator under a static transversal drive. The periodicity of the evolution in the quantum Rabi model is reproduced up to a scaling factor.

By applying the unitary displacement transformation

$$\hat{D} = \exp\left\{-\frac{\eta_r}{2\omega_{\text{eff}}}\left(\hat{b}^\dagger - \hat{b}\right)\right\} \tag{6.22}$$

the additional term translates into a qubit tunneling term $\propto \hat{\sigma}_x$, giving rise to a sub-rotation of the effective frame. We obtain the quantum Rabi Hamiltonian in its original form

$$\begin{aligned}\hat{D}^\dagger \frac{\hat{H}_{\text{eff,p}}}{\hbar}\hat{D} &= \frac{\eta_2}{2}\frac{\hat{\sigma}_z}{2} + \omega_{\text{eff}}\left(\hat{b}^\dagger - \frac{\eta_r}{2\omega_{\text{eff}}}\right)\left(\hat{b} - \frac{\eta_r}{2\omega_{\text{eff}}}\right)\\ &+ \frac{g}{2}\sigma_x\left(\hat{b}^\dagger + \hat{b} - \frac{\eta_r}{\omega_{\text{eff}}}\right) + \frac{\eta_r}{2}\left(\hat{b}^\dagger + \hat{b} - \frac{\eta_r}{2\omega_{\text{eff}}}\right) \quad (6.23)\\ &= \frac{\eta_2}{2}\frac{\hat{\sigma}_z}{2} - g\frac{\eta_r}{2\omega_{\text{eff}}}\hat{\sigma}_x + \omega_{\text{eff}}\hat{b}^\dagger\hat{b} + \frac{g}{2}\sigma_x\left(\hat{b}^\dagger + \hat{b}\right) + \text{const.}, \quad (6.24)\end{aligned}$$

using

$$\hat{D}^\dagger\hat{b}^\dagger\hat{b}\hat{D} = \hat{D}^\dagger\hat{b}^\dagger\hat{D}\hat{D}^\dagger\hat{b}\hat{D} = \left(\hat{b}^\dagger - \frac{\eta_r}{2\omega_{\text{eff}}}\right)\left(\hat{b} - \frac{\eta_r}{2\omega_{\text{eff}}}\right). \tag{6.25}$$

The resulting dynamics complies with the envelope defined by the ideal Hamiltonian with the tunneling term absent and therefore maps to the ideal quantum Rabi model, leaving its dynamics qualitatively unaffected. Master equation simulations supporting these statements are depicted in Fig. 6.8(a), (b). The thick black line in Fig. 6.8(a) shows a master equation simulation of the qubit population according to the effective Hamiltonian, Eq. (6.12), for $\eta_r = 0$. When switching on the parasitic driving, $\eta_r \neq 0$, the qubit is subject to a sub-rotation that however adheres with the envelope defined by the pure Hamiltonian. The dynamics of the bosonic mode remains unchanged by the parasitic drive, as the displacement transformation defined in Eq. (6.22) leaves the eigenenergies of the isolated harmonic oscillator unchanged,

$$\hat{D}^\dagger\left(\omega_{\text{eff}}\hat{b}^\dagger\hat{b} + \frac{1}{2}\eta_r(\hat{b}^\dagger + \hat{b})\right)\hat{D} = \omega_{\text{eff}}\hat{b}^\dagger\hat{b} + \text{const.} \tag{6.26}$$

The time evolution of the driven harmonic oscillator Hamiltonian in the rotating frame,

$$\frac{\hat{H}_h}{\hbar} = \omega_{\text{eff}}\hat{b}^\dagger\hat{b} + \frac{1}{2}\eta_r(\hat{b}^\dagger + \hat{b}) \tag{6.27}$$

is depicted in Fig. 6.8(b) for varying η_r.

The qubit tunneling term emerging in Eq. (6.24) appears as a correction of the dominant Rabi frequency η_1 after the transformation into the ω_1-rotating frame, see Eq. (6.6). The resulting effective Rabi frequency becomes

$$\eta_{1,\text{eff}} = \eta_1 - g\frac{\eta_r}{2\omega_{\text{eff}}} \tag{6.28}$$

and the transformation into the interaction picture, Eq. (6.11), is performed with respect to $\eta_{1,\text{eff}}$. This amounts to an effective decrease of the dominant Rabi frequency, which can be observed in the laboratory frame dynamics by a decrease in the fast oscillation frequency of the qubit population. We find quantitative agreement between the analytical expectation in Eq. (6.28) and the observed oscillation frequencies in master equation simulations of the driven laboratory frame Hamiltonian with and without parasitic drive term, see Sec. 6.8.

For a realistic parasitic coupling $\eta_r \sim 0.1\eta_1$ and $\omega_{\text{eff}}/2\pi = 5\,\text{MHz}$, we obtain an effective qubit tunneling term

$$-g\frac{\eta_r}{2\omega_{\text{eff}}}\hat{\sigma}_x \sim -2\pi \times 2.2\,\text{MHz} \times \hat{\sigma}_x \sim -0.4\omega_{\text{eff}} \times \hat{\sigma}_x. \tag{6.29}$$

The effective change in the dominant Rabi frequency is on the order of 5%. Therewith, only a small modification to the frequency constraint $\omega_1 - \omega_2 = \eta_{1,\text{eff}}$ is necessary due to the parasitic driving of the bosonic mode.

6.7.3 Phase dependent qubit response

While the phase of the qubit Bloch vector is not well defined for initial states $|g\rangle$, $|e\rangle$, the qubit state carries phase information when prepared on the equatorial plane of the Bloch sphere via a $\pi/2$ pulse. Figure 6.9 shows the qubit time evolution with varying relative phase φ_1. The plots are given in the original qubit basis, as calibrated in a Rabi oscillation experiment.

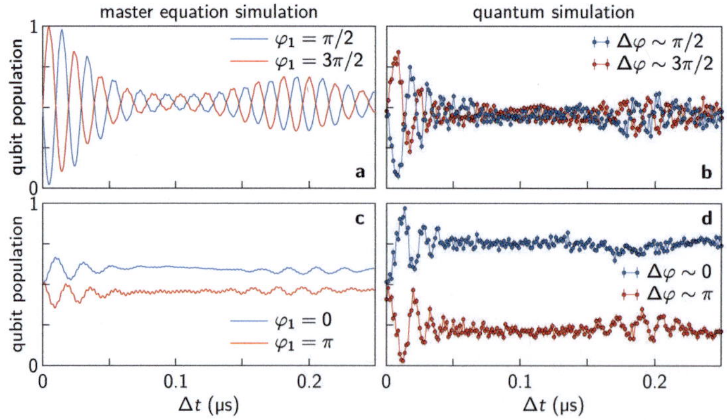

Figure 6.9: Qubit time evolution for a varying relative phase φ_1 of the applied drive The initial qubit state is prepared on the equator of the Bloch sphere $|g\rangle \pm |e\rangle$. Dispersive shifts induced by the bosonic mode are subtracted based on its population evolution as obtained in a master equation simulation.

Experimentally, the orientation of the coordinate system is set by the first microwave pulse and we apply the Rabi drive with a varying relative phase φ_1, corresponding to the angle between qubit Bloch vector and rotation axis of the drive in the equatorial plane. When both are perpendicular, $\varphi_1 = \pm\pi/2$ in Fig. 6.9(a), (b), we record a similar dynamics as observed in Fig. 6.5. The steady state at long simulation times is in the equatorial plane. For the case where $\varphi_1 = 0, \pi$ in Fig. 6.9(c), (d), qubit oscillations in the laboratory frame are suppressed while the baseline is shifted up or down, dependent on the choice of φ_1. The remaining substructure emerges from the swap interaction term between qubit and bosonic mode in the laboratory frame Hamiltonian and may be regarded as a perturbation since $\eta_1 \gg g$.

Master equation simulations confirm that the basis shift, dependent on the prepared initial qubit state, is enhanced by the presence of the second excited transmon level and by a spectral broadening of the applied Rabi drive. The experimentally observed shift is not entirely captured by the master equation simulation which we attribute to missing terms in the master equation that may be related to qubit tuning pulses and are unknown at present. We conjecture that the basis shifts are caused by an effective tilt of the qubit Bloch sphere as an artefact of the frequency tuning in experiment prior to applying the Rabi drives. We isolate the effect as an initialization issue since the basis shift cancels out in good approximation when averaging the mutually anti-parallel simulation sequences in Fig. 6.9. Dependent on φ_1, we observe a varying maximum photon population of the bosonic mode in master equation simulations. This is also indicated in raw data of the measured dispersive shift of the readout resonator.

The qubit population as depicted in Fig. 6.9 is retrieved from measured raw data by subtracting the contribution of the bosonic mode as described in Sec. 6.7.1. A deviation of the effective qubit basis is likewise observed for preparing the qubit in one of its eigenstates $|g\rangle$, $|e\rangle$, see Fig. 6.5, 6.6. The calibration of the original qubit basis is presented in the Supplementary Material of Ref. [Bra+17].

6.8 Simulation of the full quantum Rabi model

In order to simulate the full quantum Rabi model including a non-vanishing qubit energy term we switch on the second drive, $\eta_2 \neq 0$, see the pulse sequence in Fig. 6.10(a). Quantum simulations are performed with the qubit prepared in $|g\rangle$, subject to a thermal excess population. For the simulation scheme to be valid, we need to fulfil the parameter constraint $\omega_2 = \omega_1 - \eta_1$, according to the schematics in Fig. 6.10(b). We initially record a simulation sequence with $\eta_2 = 0$ and extract the Rabi frequency η_1 from a fit to the Fourier transformation of the qubit time evolution. Subsequently, we apply the same sequence including the second drive, $\eta_2 > 0$ and ω_2 based on the extracted value for η_1 in order to obey the constraint $\omega_2 = \omega_1 - \eta_1$. By sweeping the relative phase $\varphi_1 - \varphi_2$ between the two drives we calibrate the condition where $\varphi_1 = \varphi_2$.

Figure 6.10(c) shows a master equation simulation of the complete quantum Rabi model for $\eta_2 = 0$ (grey) and $\eta_2 \neq 0$ (red), respectively. The main difference is an emerging substructure between quantum revivals and a slight increase of the revival amplitude in the presence of the qubit energy term. In measured data, see Fig. 6.10(d), the additional substructure could not be fully reproduced. We attribute this to pulse imperfections caused by ring up dynamics of the rectangular drive pulses. The frequency of the short pulses is thereby not well defined such that the frequency constraint of the simulation scheme is not sufficiently well satisfied. In particular, the frequency beating in measured data leads to a broadening of the Rabi frequency η_1 in Fourier space. One can observe in Fig. 6.10(d) that more pronounced oscillations are better visible at regions of smaller oscillation frequency, namely during the collapse and around the revival. An increased revival amplitude is likewise indicated.

In principle, pulses of higher fidelity can be achieved in experiment by using Gaussian shaped envelopes, since high frequency Fourier components are suppressed. This is however incompatible with the simulation scheme, which requires rectangular pulses of constant Rabi frequency. These limitations of the simulation scheme become visible in particular in the presented experiment including two Rabi drives.

In addition, the parasitic drive of the bosonic mode contributes to a suppression of the expected signatures, which is suggested by master equation simulations. In our experimental scheme, we in fact extract the effective Rabi frequency $\eta_{1,\text{eff}}$ from Fourier transformation, since the shift induced by the parasitic drive is included in the fast oscillation frequency, as discussed in Sec. 6.7.2. The frequency constraint is therefore correctly calibrated. The change in the effective Rabi frequency $\eta_{1,\text{eff}}$ can be observed by comparing the frequencies of the red lines in Fig. 6.10(c), (e). Figure 6.10(e) shows a master equation simulation in the presence of parasitic driving, which leads to a decrease in the Rabi frequency. The frequency shift is indicated by arrows close to $\Delta t = 0.15\,\mu\text{s}$ as a guide to the eye.

By performing master equation simulations with a varying parasitic drive strength, we note a qualitative dependence of the bosonic mode population. We conjecture that this is due to the induced sub-rotation discussed in Sec. 6.7.2 and similar to the observations in Sec. 6.7.3 for varying the initial qubit phase.

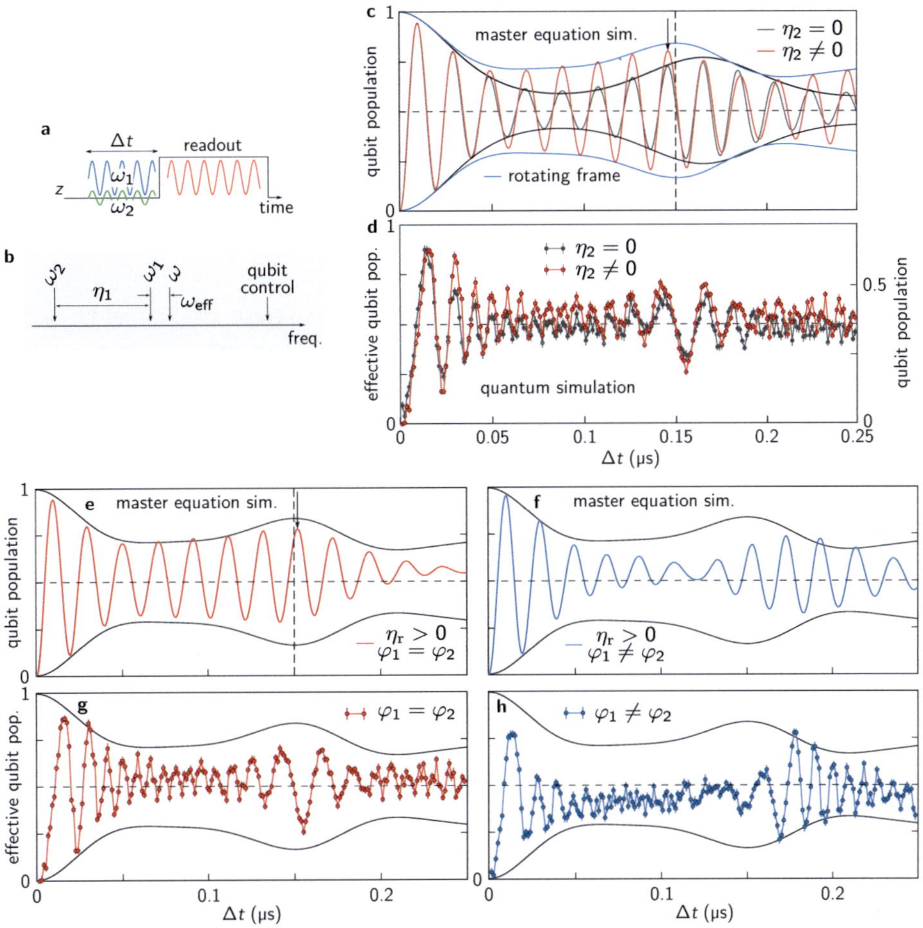

Figure 6.10: Simulation of the full quantum Rabi model (a) Schematic pulse sequence used in the experiment. (b) Overview on the relative frequencies of the bosonic mode and the applied drives. The constraint $\eta_1 = \omega_1 - \omega_2$ is sketched. (c) Master equation simulations for a vanishing qubit term $\eta_2 = 0$ (grey) and with non-vanishing qubit term $\eta_2 > 0$ (red). The blue line corresponds to the master equation simulation in the effective frame for $\eta_2/2\pi = 3\,\text{MHz}$. (d) Quantum simulation for equal parameters. The dispersive shift of the readout resonator induced by the bosonic mode is subtracted based on a master equation simulation and we added a small offset between the traces for better visibility. (e)-(h) Dependence on the relative phase $\varphi_1 - \varphi_2$ in the presence of a parasitic drive, $\eta_r = 0.05\eta_1$. Panels (e) and (f) show master equation simulations for $\varphi_1 = \varphi_2$ (e) and $\varphi_1 \neq \varphi_2$ (f). The respective experimental traces are given in (g), (h). The trends are reproduced in measured data, in particular with respect to the position of the first revival and the oscillation amplitude before the revival. The decrease in the Rabi frequency due to the parasitic drive is indicated in (c) and (e) by arrows.

In order to further demonstrate the effect of the second Rabi drive, we sweep its relative phase $\varphi_1 - \varphi_2$. Figures 6.10(e), (f) show master equation simulations including the parasitic coupling to the bosonic mode, while the associated experimental traces are depicted in Fig. 6.10(g), (h), respectively. While deviations between theory and experiment are present, we observe the trend that the revival is shifted to later times for a violated phase condition $\varphi_1 \neq \varphi_2$ and the oscillation amplitude before the first revival is suppressed. The correct scenario is depicted in Fig. 6.10(g), being the identical data set as in Fig. 6.10(d).

We estimate $\eta_2/2\pi \sim 3\,\mathrm{MHz}$ by comparing the relative peak heights of both drive tones with a spectrum analyser. With $\omega_{\mathrm{eff}}/2\pi = 6\,\mathrm{MHz}$ we approach a regime where $2g_{\mathrm{eff}}/\sqrt{\omega_{\mathrm{eff}}\eta_2/2} > 1$, marking the quantum critical point in the related Dicke model [NC10].

6.9 Efficient generation of non-classical cavity states

The presented simulation scheme provides an efficient technique to generate non-classical states in the Fock space of the bosonic mode. We demonstrate this in Fig. 6.11 by plotting the time evolution of the Wigner function $W(X, P)$ of the bosonic mode after post-selecting the qubit in either of its basis states. We use the Wigner function representation

$$W(X, P, t) = \frac{1}{\pi\hbar}\int_{-\infty}^{\infty} \mathrm{d}Y \psi^*(X + Y, t)\psi(X - Y, t)e^{2\mathrm{i}PY}, \tag{6.30}$$

where $X \propto \hat{b}^\dagger + \hat{b}$ and $P \propto \mathrm{i}(\hat{b}^\dagger - \hat{b})$ refer to the conjugate variables of the bosonic mode and $\psi(X, t)$ denotes the pure state wavefunction of the bosonic mode after an evolution time t.

Post-selection is performed numerically by projecting the evolved state $|\Psi(t)\rangle$ of the joint Hilbert space according to

$$\begin{aligned} |\Psi(t)\rangle_{|\mathrm{g}\rangle} &= (\mathbb{1}_\mathrm{b} \otimes |\mathrm{g}\rangle\langle\mathrm{g}|)\,|\Psi(t)\rangle \\ |\Psi(t)\rangle_{|\mathrm{e}\rangle} &= (\mathbb{1}_\mathrm{b} \otimes |\mathrm{e}\rangle\langle\mathrm{e}|)\,|\Psi(t)\rangle . \end{aligned} \tag{6.31}$$

The Wigner function is evaluated using the bosonic mode state $|\psi(t)\rangle$. It is obtained by taking the partial trace in the qubit Hilbert space,

$$\begin{aligned} |\psi(t)\rangle_{|\mathrm{g}\rangle} &= \mathrm{tr}_\mathrm{q}\,|\Psi(t)\rangle_{|\mathrm{g}\rangle} \\ |\psi(t)\rangle_{|\mathrm{e}\rangle} &= \mathrm{tr}_\mathrm{q}\,|\Psi(t)\rangle_{|\mathrm{e}\rangle} . \end{aligned} \tag{6.32}$$

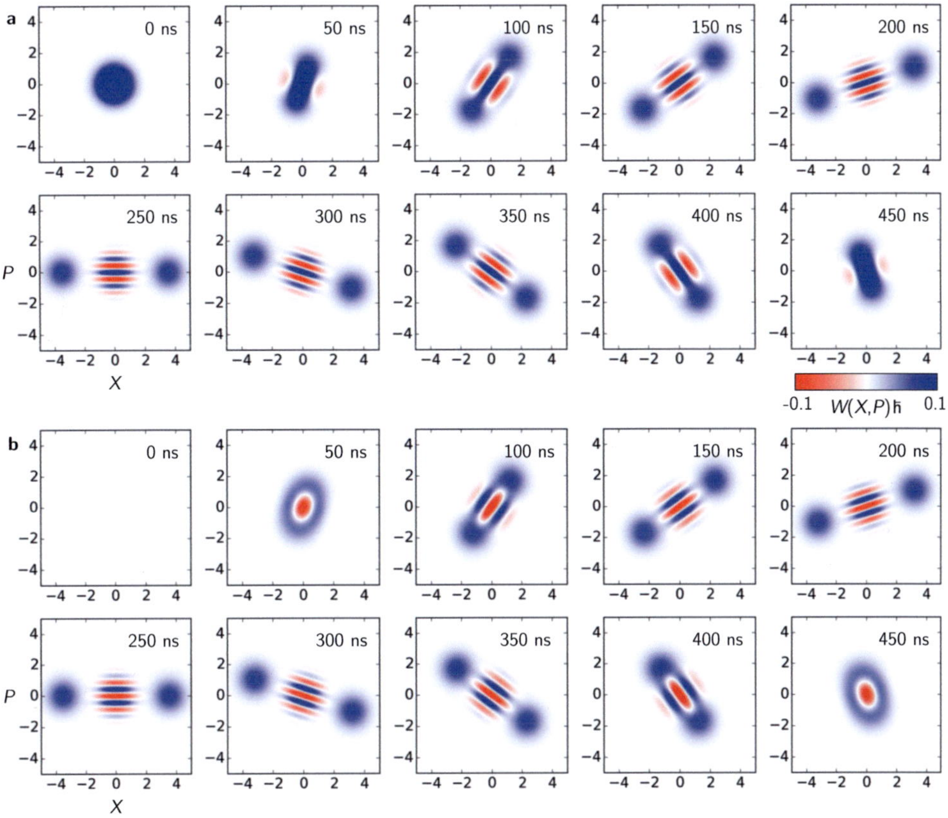

Figure 6.11: Wigner function evolution of the bosonic mode for the quantum Rabi Hamiltonian at USC conditions The plots show the time evolution of the Wigner function $W(X, P)$ in the absence of dissipation as calculated from a master equation simulation. The qubit state is post-selected either in state $|g\rangle$ (a) or $|e\rangle$ (b). Depending on the qubit state one can observe the formation of even $|\alpha\rangle + |-\alpha\rangle$ (a) and odd $|\alpha\rangle - |-\alpha\rangle$ (b) cat states, respectively, which differ in their unique interference patterns. $|\alpha\rangle$ denotes a coherent state. $X \propto \hat{b}^\dagger + \hat{b}$ and $P \propto \mathrm{i}(\hat{b}^\dagger - \hat{b})$ refer to the conjugate variables of the bosonic mode.

Experimentally, the post-selection requires a single-shot qubit readout that is feasible with parametric quantum amplifiers. This allows for a correct allocation of the measured resonator state according to the qubit state. In order to perform a full Wigner tomography protocol, projection measurements with the help of an ancillary qubit need to be repeated multiple times in order to record every pixel of the Wigner plot. The mentioned experimental techniques are sophisticated and were not available when the above experiment was performed. However, the simulations demonstrate the eligibility of the presented simulation scheme for fast and efficient creation of non-classical cavity states.

The master equation simulation underlying Fig. 6.11 was performed for the effective quantum Rabi Hamiltonian at USC conditions in the absence of a qubit term, see Eq. (6.16). Simulation parameters orientate at the presented experiment with $\omega_{\text{eff}}/2\pi = 2\,\text{MHz}$ and $g/2\pi = 5\,\text{MHz}$ and we neglected dissipation.

6.10 Conclusion

We have demonstrated an analog quantum simulation of the quantum Rabi model in the ultra-strong coupling regime. A quantum state collapse and revival as the distinct signatures of this regime were observed in the qubit dynamics, validating the experimental feasibility of the proposed simulation scheme [Bal+12]. While the accessible coupling regime is fundamentally not limited by the scheme, we could experimentally observe quantum revivals up to a coupling regime where $g_{\text{eff}}/\omega_{\text{eff}} \sim 0.6$ due to the finite coherence and unwanted cross-talk in our circuit. A limitation of the scheme is an effective slowing down of the system dynamics, while the laboratory frame dissipation rates are maintained in the synthesized frame. In analogy to the measure of cooperativity in standard QED, we find the ratio $g_{\text{eff}}/\sqrt{\kappa/T_1} \sim 30$, rendering the qubit and bosonic mode decay rates an ultimate limitation for the simulation quality. The decelerated system dynamics in the effective frame however allows for the observation of quantum revivals on a time-scale of $\sim 100\,\text{ns}$, while the repetition time of revivals in the laboratory frame USC quantum Rabi model is on a time scale of $2\pi\omega^{-1} \sim 0.2\,\text{ns}$, being experimentally hard to resolve. A major limitation of the simulation scheme emerged in simulating the full quantum Rabi model with two Rabi drives applied. Presumably due to the rectangular shape of the drive pulses, as required by the simulation scheme, we observed a frequency beating in the experimental simulations. As a consequence, the parameter constraints required by the scheme were satisfied only to a limited extend, which explains the encountered deviations from master equation simulations. The parasitic coupling of the dominant Rabi drive to the bosonic mode further limits the simulation quality.

7 Towards quantum simulation of the spin boson model

By substituting the single harmonic oscillator mode of the quantum Rabi model by a set of oscillator modes that form a bosonic bath, we can engineer the spin boson model. Similar to the quantum Rabi model, the spin boson model is described by an infinitely large Hilbert space and the model exhibits non-classical ground state properties and dynamics in certain parameter regimes. The spin boson model can capture basic biological phenomena in a more realistic way than the simplified quantum Rabi model investigated in the previous chapter.

In principle one can integrate the spin boson model numerically by truncating the Hilbert spaces of the resonator modes at a large enough maximum Fock number in order to recover a finite problem. However, it turns out that numerical solutions are very sensitive on the choice of the truncation, such that they regularly diverge [FG96]. Several numerical methods have been devised to study the dynamics of the spin boson model, such as the non-interacting blip approximation (NIBA) [Leg+87], non-Markovian master equations [Koc+08], Monte Carlo methods [Gul+11], renormalization group approaches [AS06] and several others [OIL13]. For specific parameter regimes, the results of these techniques have been shown to be of questionable validity and are not always numerically exact [ABV07; OIL13]. The vision of the present approach is to experimentally quantum simulate the dynamics of the spin boson model in various regimes and thereby provide indications which of the presented solutions obtained in numerical approaches are most promising.

In this chapter we present an on-chip implementation of the spin boson model based on a modular flip-chip approach and show experimental evidence that the dissipative qubit dynamics is governed by the engineered bosonic bath. This can be a starting point to observe a variety of interesting phenomena in the spin boson model [Lep+17].

7.1 The spin boson model

The spin boson model generically describes the effect of a dissipative environment on the dynamics of a quantum mechanical two-level system, or qubit [Leg+87; FVN10]. While the dynamics of an isolated qubit can be obtained analytically, solving the complete spin boson model becomes non-trivial for an open quantum system in an arbitrary environment. In order to access and exploit the quantum properties of a qubit or a more complex quantum circuit, a finite interaction with its environment is required.

The quantum impurity couples to the electromagnetic field of the bosonic environment via its electric dipole moment. Provided that the interaction strength is sufficiently weak, the environment can be represented as a set of harmonic oscillators with a coupling strength that is linear in the oscillator coordinates. In addition, it was argued that it is adequate and sufficiently general to treat the effect of the environment on the qubit with an interaction term only proportional to one qubit degree of freedom, say $\hat{\sigma}_x$ [Leg+87].

The general spin boson Hamiltonian, a variant of the Caldeira-Legget model [CL83], reads

$$\frac{\hat{H}}{\hbar} = \frac{\epsilon}{2}\hat{\sigma}_z + \frac{\Delta}{2}\hat{\sigma}_x + \sum_i \omega_i \hat{b}_i^\dagger \hat{b}_i + \hat{\sigma}_x \sum_i g_i \left(\hat{b}_i^\dagger + \hat{b}_i\right), \tag{7.1}$$

with ϵ the qubit transition frequency, Δ a qubit tunneling matrix element, ω_i the frequencies of the oscillator modes and g_i the individual coupling strengths between the qubit and bosonic mode i. $\hat{\sigma}_i$ are Pauli matrices with $\hat{\sigma}_z \left|\mathrm{g}\right\rangle = -\left|\mathrm{g}\right\rangle$ and $\hat{\sigma}_z \left|\mathrm{e}\right\rangle = \left|\mathrm{e}\right\rangle$, where $\left|\mathrm{g}\right\rangle$, $\left|\mathrm{e}\right\rangle$ denote the eigenstates of the computational qubit basis. $\hat{b}_i^\dagger$ ($\hat{b}_i$) are creation (annihilation) operators in the Fock space of the i-th oscillator mode. In quantizing the harmonic oscillator, we obtain bosonic commutation relations

$$\left[\hat{b}_i, \hat{b}_j^\dagger\right] = \hat{b}_i\hat{b}_j^\dagger - \hat{b}_j^\dagger\hat{b}_i = \delta_{ij},\ \delta_{ij} = \begin{cases} 1 & \text{if } i = j \\ 0 & \text{if } i \neq j \end{cases}, \tag{7.2}$$

with δ_{ij} the Kronecker delta. The oscillator modes are therefore referred to as bosonic modes and we call the collective of the bosonic modes the bosonic bath. The fact that the qubit Hamiltonian represents a particle with spin-1/2 in a constant and parallel magnetic field, motivates the name spin boson model.

The spin boson Hamiltonian (7.1) becomes non-diagonal and therefore non-trivial for $\epsilon \neq 0$, rendering it an appealing testbed for quantum simulations. Since the effects of the tunneling term are negligible in many scenarios of practical interest, we focus on the case where $\Delta = 0$.

The notation in Eq. (7.1) is equivalent to the literature notation, where the $\hat{\sigma}_z$-operator typically appears in the coupling term and the $\hat{\sigma}_x$-operator in the qubit term. Our notation is motivated by considering the spin boson model as a many-mode extension of the quantum Rabi model. Besides, it allows for a direct mapping onto a cQED system involving a single qubit, transversally coupled to a set of harmonic oscillator modes. For the transmon qubit in particular, the tunneling matrix element vanishes, since hopping between wells in the Josephson potential are suppressed due to its periodicity. For a small inductive term in the transmon Hamiltonian this argument remains valid in the spirit of the WKB approximation.

7.2 Bath spectral density $J(\omega)$

In thermal equilibrium, the properties of the bosonic bath as experienced by the qubit are encapsulated in the bath spectral density $J(\omega)$ [Leg+87; VEA12], defined as

$$J(\omega) \equiv \pi\hbar \sum_i g_i^2 \delta(\omega - \omega_i). \tag{7.3}$$

By considering only positive bosonic mode frequencies $\omega_i > 0$, $J(\omega)$ is non-zero only for $\omega > 0$. By assuming that $J(\omega)$ is a smooth function and can be written as a power law up to a cut-off frequency ω_c, we obtain

$$J(\omega) = \frac{\pi}{2}\alpha\hbar\omega^s \omega_c^{-s+1} \mathfrak{C}(\omega, \omega_c). \tag{7.4}$$

α denotes the dimensionless Kondo parameter, which is interpreted as a dissipation strength [SMS02] and $\mathfrak{C}(\omega, \omega_c)$ is a cut-off function.

The exponent s describes the shape of the spectral function. $s = 1$ corresponds to a linear spectral function, also called ohmic, where $J(\omega) \propto \omega$, as is the case for a $50\,\Omega$ matched transmission line. The regime where $s < 1$ is called sub-ohmic and turns out to be particularly appealing to study in the framework of quantum simulations, since numerical exact solutions are scarce.

The bath spectral density, as defined in Eq. (7.3), corresponds to the density of states

$$\rho(\omega) = \frac{1}{\hbar} \sum_i \delta(\omega - \omega_i), \tag{7.5}$$

with single mode contributions weighted by the square of the individual coupling strengths g_i. The effect of the bosonic environment on the qubit is entirely captured by the bath spectral density $J(\omega)$ [FVN10], since any correlator of more than two bath operators can be split into a product of correlators of only two bath operators. This is due to the underlying Gaussian statistics of the bosonic modes [Lep+17].

7.3 Bath spectral function $S(\omega)$

In order to account for a finite temperature, we introduce the spectral function $S(\omega)$. It is defined as the Fourier transform of the bath correlator [VEA12],

$$S(\omega) = \int_{-\infty}^{\infty} \mathrm{d}\tau e^{\mathrm{i}\omega\tau} \langle \hat{X}(\tau)\hat{X}(0)\rangle, \tag{7.6}$$

with the bath operator [Hau15; Paz+16]

$$\hat{X} = \sum_i \hbar g_i \left(\hat{b}_i^\dagger + \hat{b}_i\right). \tag{7.7}$$

Sometimes, only the symmetrized version of the bath correlator $\langle[\hat{X}(\tau), \hat{X}(0)]_+\rangle$ enters the definition of $S(\omega)$, referred to as the classical component of the spectral function [Paz+16]. From

$$\langle[\hat{X}(\tau), \hat{X}(0)]_+\rangle + \langle[\hat{X}(\tau), \hat{X}(0)]_-\rangle = \langle\hat{X}(\tau)\hat{X}(0)\rangle, \tag{7.8}$$

we can see that the complete form of $S(\omega)$ is obtained from the sum of the anti-commutator $[,]_+$ and the commutator $[,]_-$ representation of the bath operator products.

We can evaluate Eq. (7.6) by inserting the definition of $\hat{X}$ in Eq. (7.7),

$$\begin{aligned}
S(\omega) &= \int_{-\infty}^{\infty} \mathrm{d}\tau e^{\mathrm{i}\omega\tau} \sum_{i,j} \hbar^2 g_i g_j \left\langle \left(\hat{b}_i^\dagger(\tau) + \hat{b}_i(\tau)\right)\left(\hat{b}_j^\dagger + \hat{b}_j\right)\right\rangle && (7.9)\\
&= \hbar^2 \sum_i g_i^2 \int_{-\infty}^{\infty} \mathrm{d}\tau e^{\mathrm{i}\omega\tau} \left[\langle\hat{b}_i^\dagger\hat{b}_i\rangle e^{\mathrm{i}\omega_i\tau} + \langle\hat{b}_i\hat{b}_i^\dagger\rangle e^{-\mathrm{i}\omega_i\tau}\right] && (7.10)\\
&= \hbar^2 \sum_i g_i^2 \int_{-\infty}^{\infty} \mathrm{d}\bar{\omega}\delta(\bar{\omega} - \omega_i) \int_{-\infty}^{\infty} \mathrm{d}\tau \left[n(\bar{\omega})e^{\mathrm{i}\bar{\omega}\tau} + (1 + n(\bar{\omega}))e^{-\mathrm{i}\bar{\omega}\tau}\right] && (7.11)\\
&= \frac{\hbar}{\pi} \int_0^{\infty} \mathrm{d}\bar{\omega} J(\bar{\omega}) \int_{-\infty}^{\infty} \mathrm{d}\tau e^{\mathrm{i}\omega\tau} \left[n(\bar{\omega})e^{\mathrm{i}\bar{\omega}\tau} + (1 + n(\bar{\omega}))e^{-\mathrm{i}\bar{\omega}\tau}\right]\\
&= \frac{\hbar}{\pi} \int_{-\infty}^{\infty} \mathrm{d}\tau e^{\mathrm{i}\omega\tau} \left[\int_{-\infty}^{0} \mathrm{d}\bar{\omega} J(-\bar{\omega}) n(-\bar{\omega}) e^{-\mathrm{i}\bar{\omega}\tau} + \int_0^{\infty} \mathrm{d}\bar{\omega} J(\bar{\omega})(1 + n(\bar{\omega}))e^{-\mathrm{i}\bar{\omega}\tau}\right]\\
&= \frac{\hbar}{\pi} \int_{-\infty}^{\infty} \mathrm{d}\tau e^{\mathrm{i}\omega\tau} \left[\int_{-\infty}^{\infty} \mathrm{d}\bar{\omega} J(|\bar{\omega}|)\mathrm{sgn}(\bar{\omega})(1 + n(\bar{\omega}))e^{-\mathrm{i}\bar{\omega}\tau}\right] && (7.12)\\
&= 2\hbar J(|\omega|)\mathrm{sgn}(\omega)(1 + n(\omega))\\
&= \hbar J(|\omega|)\mathrm{sgn}(\omega)\left(1 + \coth\left(\frac{\beta\hbar\omega}{2}\right)\right). && (7.13)
\end{aligned}$$

In order to arrive at Eq. (7.10), we employed the time evolution operator $\hat{U}(\tau) = e^{-i\hat{H}_b\tau/\hbar}$ of the bosonic bath Hamiltonian $\hat{H}_b = \sum_i \hbar\omega_i \hat{b}_i^\dagger \hat{b}_i$, and the identities

$$\hat{b}_i^\dagger(\tau) = \hat{U}^\dagger(\tau)\hat{b}_i^\dagger\hat{U}(\tau) = e^{i\hat{H}_b\tau}\hat{b}_i^\dagger e^{-i\hat{H}_b\tau} = \hat{b}_i^\dagger e^{i\omega_i\tau} \tag{7.14}$$

$$\hat{b}_i(\tau) = \hat{U}^\dagger(\tau)\hat{b}_i\hat{U}(\tau) = e^{i\hat{H}_b\tau}\hat{b}_i e^{-i\hat{H}_b\tau} = \hat{b}_i e^{-i\omega_i\tau}. \tag{7.15}$$

Additionally, we introduced the Bose-Einstein statistics

$$n(\omega) = \frac{1}{e^{\beta\hbar\omega} - 1} = \frac{1}{2}\left(\coth\left(\frac{\beta\hbar\omega}{2}\right) - 1\right) \tag{7.16}$$

in Eq. (7.11), with $\beta^{-1} \equiv k_B T$. In order to account for negative frequencies ω we use the relation $n(-\omega) = -(n(\omega)+1)$ and the sign function $\mathrm{sgn}(\omega)$ to arrive at Eq. (7.12).

The close relation between the spectral function $S(\omega)$ and the bath spectral density $J(\omega)$ is revealed most clearly at $T = 0$,

$$S(\omega)|_{T=0} = 2\hbar J(|\omega|)\mathrm{sgn}(\omega). \tag{7.17}$$

Likewise, $S(\omega)$ agrees with $J(\omega)$ up to a constant factor for $\omega \gg T$, while $S(\omega = 0) \propto T$ for an ohmic spectral density.

7.3.1 $1/f$ noise

As an example, the spectral function reduces to

$$S(\omega) = \pi\hbar^2\alpha\omega_c\mathfrak{C}(\omega,\omega_c)\coth\left(\frac{\beta\hbar\omega}{2}\right) \tag{7.18}$$

in the case where $s = 0$ in the bath spectral density, Eq. (7.4). In the limit $\beta\hbar\omega \ll 1$, we can expand $\coth(\beta\hbar\omega/2) \approx 2(\beta\hbar\omega)^{-1}$ and Eq. (7.18) becomes

$$S(\omega \ll \beta^{-1}) = \pi\hbar\alpha\omega_c\mathfrak{C}(\omega,\omega_c)\frac{2}{\beta\omega}, \tag{7.19}$$

yielding $1/f$ noise that is incessantly present in experiment and in general of unclear origin.

7.3.2 Relation to a circuit impedance

From the fluctuation dissipation theorem, we can derive a formula connecting the spectral function $S(\omega)$ of a bosonic environment to its equivalent impedance $Z(\omega)$ [SMS02; Lep+17],

$$S(\omega)|_{T=0} \propto \omega\Re Z(\omega). \tag{7.20}$$

The proportionality factor contains information on how voltage fluctuations in the environment characterized by $Z(\omega)$ influence the qubit. It is apparent from this relation that a 50 Ω matched transmission line with $Z(\omega) = \text{const}$ yields an ohmic spectral function with a linear frequency dependence as pointed out in Sec. 7.2,

$$S(\omega) \propto \omega. \tag{7.21}$$

For completeness, we note that $S(\omega)$ corresponds to the expected Lorentzian distribution for a parallel LCR resonator. With lumped-element resistance R, capacitance C, inductance L, and internal quality factor Q_i, as defined in Ch. 3, we obtain [Poz12]

$$Z(\omega) = \left(\frac{1}{R} + \mathrm{i}\omega C + \frac{1}{\mathrm{i}\omega L}\right)^{-1} \approx \frac{R}{1 + 2\mathrm{i}Q_\mathrm{i}\Delta\omega/\omega_0} \tag{7.22}$$

and therefore

$$S(\omega)_{T=0} \propto \omega\Re Z(\omega) \propto \frac{\omega}{1 + 4Q_\mathrm{i}^2\left(\frac{\Delta\omega}{\omega_0}\right)^2} \approx \frac{\omega_0}{1 + 4Q_\mathrm{i}^2\left(\frac{\Delta\omega}{\omega_0}\right)^2}. \tag{7.23}$$

7.4 Experimental concept

It is our goal to engineer a tailored bosonic bath that influences the dissipative qubit dynamics in a controlled way. We want to achieve a scenario where the qubit dissipation into the bosonic bath is dominating the total decay rate of the qubit. As a consequence, the effect of unwanted and possibly unknown decay channels can be neglected.

Figure 7.1(a) shows the generic circuit diagram of a transmon qubit coupled to an environment characterized by the equivalent bosonic bath impedance $Z_\mathrm{b}(\omega)$. In our experiment, we use a frequency tunable concentric transmon qubit with identical geometric parameters as detailed in Ch. 5. At a fundamental qubit frequency $\epsilon/2\pi = 7.412\,\mathrm{GHz}$ we measure a transmon anharmonicity of $\alpha/h = -0.30\,\mathrm{GHz}$. The relevant capacitance accounting for the qubit is $C_\mathrm{int} = C_\mathrm{J}C_\mathrm{g}/(C_\mathrm{J} + C_\mathrm{g})$, given as the series capacitance of C_J and C_g, see Fig. 7.1(b). The effective impedance $Z(\omega)$ that maps to the spectral function $S(\omega)$ therewith becomes

$$Z(\omega)^{-1} = \mathrm{i}\omega C_\mathrm{int} + Z_\mathrm{b}(\omega)^{-1}. \tag{7.24}$$

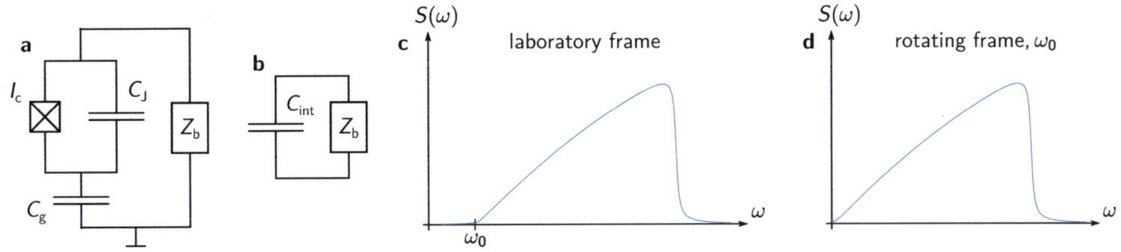

Figure 7.1: Generic spin boson circuit diagram and schematic principle of our approach (a) Transmon qubit coupled to an environment with impedance $Z_b = Z_b(\omega)$. (b) The effective qubit capacitance as seen by the bath becomes C_{int}. (c) Spectral function $S(\omega)$ of a bosonic bath with restricted frequency range in the laboratory frame. (d) In the frame rotating at ω_r, the bosonic bath has a sharp onset at zero frequency, corresponding to the case $T = 0$. In the rotating frame, we therewith avoid a finite value of the spectral function at $\omega = 0$.

In the experiment, we engineer a bosonic bath in a restricted frequency band of about 0.5 GHz, see the schematic in Fig. 7.1(c). This allows us to effectively switch off the influence of the bosonic bath by detuning the qubit from the spectral location of the bosonic bath, for instance during state preparation and readout. In addition, we can look at the qubit-bath system in a frame rotating with the lower cut-off frequency ω_0 of the bosonic bath, see Fig. 7.1(d). In the rotating frame, we obtain a natural shape of the spectral function with $S(\omega = 0) \approx 0$. Additionally, we can create an effective spectral function with a finite zero-frequency value $S(\omega = 0) > 0$ by moving into a frame rotating at a frequency $\omega > \omega_0$. In such a way, we can choose the effective system temperature by selecting the appropriate rotating frame. In bath implementations with transmission lines, the finite temperature behaviour of the spectral function is known as the infrared cut-off problem.

7.5 Flip-chip approach

In our experimental realization of the spin boson model, we distribute the relevant elements of the spin boson model on two physically different chips. The qubit with its readout and control circuitry is fabricated on an intrinsic silicon substrate, using a single electron beam lithography step and Al evaporation as detailed in Sec. 4.1. The bosonic bath is fabricated on a separate sapphire chip in an optical lift-off lithography process. Both samples are mounted inside a specifically designed sample box on top of each other in a flip-chip fashion. The qubit sample at the bottom is mounted on the ground level of the sample box, which allows for the required bond connections to the PCB. The bosonic bath sample is mounted on top of two pedestals that are machined inside the sample box, see Fig. 7.2(a). It is flipped upside down and therefore facing the qubit chip, facilitating a coupling via electric fields.

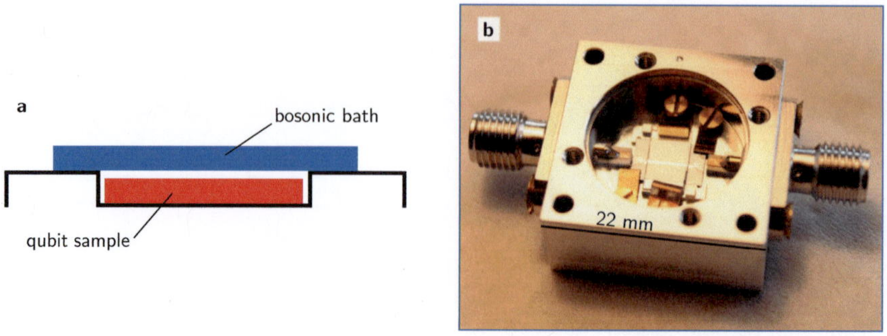

Figure 7.2: Flip-chip sample box (a) A schematic side view shows how the two samples are attached on top of each other. Their distance is adjusted by the hight of machined spacers inside the sample box. (b) Photograph of the flip-chip sample box with two samples mounted on top of each other. Samples are fixed by using CuBe clamps.

The distance between the two chips is adjusted via the height of the spacers. We designed boxes that yield a sample spacing of 35 µm to 80 µm, at a precision of ±3 µm, given by the precision of the CNC mill in our machine shop. The experiment presented in Sec. 7.7 is carried out for a sample spacing of 35 µm.

Instead of using glue that would potentially lift the samples due to the capillary effect, we hold both chips with CuBe clamps that are known to maintain their elasticity at low temperatures. See Fig. 7.2(b) for a photograph of the two samples mounted inside the box.

A main advantage of this flip-chip technique is its modularity. It allows us to measure the qubit dynamics in the presence of different bosonic baths with varying spectral functions $S(\omega)$. In addition, we can pre-characterize the qubit sample without bosonic bath attached. Moreover, fabricated bosonic bath chips can be tested by measuring microwave transmission through a separate probe sample, which is mounted as the lower chip inside an identical sample box.

7.6 Engineering of the bosonic bath

There are various approaches to engineer a specific spectral function with superconducting circuits. They have in common that the equivalent impedance $Z(\omega)$ of the bath is tailored, which is connected to its spectral function, see Eq. (7.20).

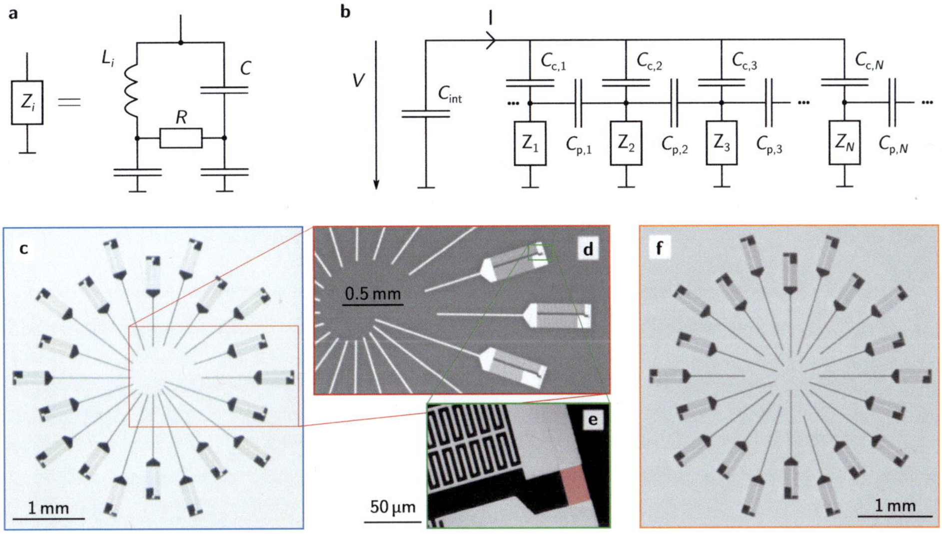

Figure 7.3: Tailored bosonic bath (a) Lumped-element circuit diagram of one of the 20 bath resonators, showing a series *LCR* circuit. (b) Circuit diagram of the bath impedance *Z* built out of many individual resonators of impedance Z_i. The bath capacitively couples to the transmon qubit which is represented by its effective capacitance C_{int}. Parasitic coupling between individual resonators is represented by capacitances $C_{\text{p},i}$. (c) Optical micrograph of a fabricated bosonic bath. The variation of the individual coupling capacitances $C_{\text{c},i}$ is visible by various lengths of the coupler arms. Resonators that are adjacent in frequency are also spatially neighbouring. (d), (e) The insets show the lumped-elements of the resonators. The resistive copper metallization is visible in red. (f) Version of the bosonic bath, with resonators that are adjacent in frequency arranged at maximum distance. The resistive element used here is an AuPd metallization, as described in the text.

The most intuitive approach to synthesize the bath Hamiltonian from Eq. (7.1) is to fabricate many individual resonators that individually couple to the qubit. In order to achieve a continuous spectral function in a fixed frequency band, the number of required resonators depends on the spectral width of the individual resonators. Since the number of resonators that can couple to a single qubit is experimentally limited, we deliberately increase their line widths.

We use 20 superconducting series lumped-element resonators that are equipped with a resistive element. With a designed line width in the range of $\sim 40\,\text{MHz}$, we can achieve a rather smooth filling of a 0.5 GHz wide frequency band. See Fig. 7.3(a) for a circuit diagram of a single resonator. The inductance L_i of resonator i is linearly varied, resulting in a linear distribution of the bare resonator frequencies in good approximation.

Table 7.1: Relevant parameters of the lumped element resonators constituting the bosonic bath We summarize the individual resonance frequencies $\omega_i/2\pi$, spectral widths $\Delta\omega_i/2\pi$, internal quality factors $Q_{\mathrm{i},i}$ and lumped-element parameters L_i, C and resonator impedance Z_i.

$\omega_i/2\pi$ (GHz)	$\Delta\omega_i/2\pi$ (MHz)	$Q_{\mathrm{i},i}$ (10^3)	L_i (nH)	C (pF)	Z_i (Ω)
7.5 .. 8.0	36 .. 40	0.21 .. 0.20	2.15 .. 1.9	0.21	101 ..95

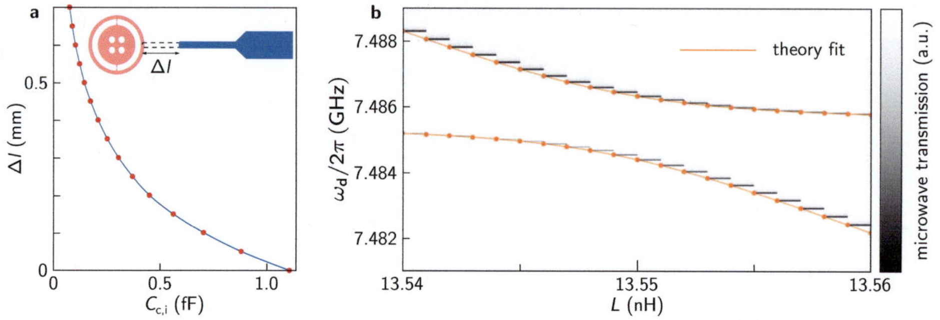

Figure 7.4: Microwave simulations for designing the bosonic bath (a) The figure shows the coupling capacitance $C_{\mathrm{c},i}$ of a single resonator of the bosonic bath to the concentric transmon qubit based on a microwave simulation. The distance between the two chips is 50 µm in the simulation. We plot the shortening Δl of the coupler arm with respect to the maximal length that yields a maximum coupling capacitance of 1.1 fF. The simulated geometry is schematically depicted in the inset. The concentric transmon (red) is located on the lower chip in the simulation, while a single resonator (blue) is on the bottom layer of the upper chip. Maximum coupling is achieved when the coupler arm touches the ring electrode of the qubit in the projection. The blue line corresponds to a spline fit of simulated data points. (b) Simulated maximum coupling strength $g_{i,\max}/2\pi = 1.0\,\mathrm{MHz}$ for $\Delta l = 0$ at a chip spacing of 35 µm.

Since the frequency band where the bath has non-zero $J(\omega)$ is small, the effective environment impedance $Z(\omega)$ must reflect the desired form of the bath spectral density. We achieve a shaping of $Z(\omega)$ by adjusting the coupling strengths g_i between qubit and the individual lumped-element resonators forming the bosonic bath. Since the coupling strengths and therefore the coupling capacitances $C_{\mathrm{c},i} \sim g_i$ appear in first order linear in $Z(\omega)$ [Lep+17], we linearly vary $C_{\mathrm{c},i} = 0.17\,\mathrm{fF}$ to $1.1\,\mathrm{fF}$ in order to recover an approximately ohmic spectral function. See Fig. 7.3(b) for a complete circuit diagram of the bosonic bath.

From microwave simulations [1], we obtain a relation between the length of the coupling antenna and the coupling capacitance $C_{\mathrm{c},i}$ between qubit and bosonic

[1] Sonnet Software Inc., Sonnet Professional 13.52

mode i. See Fig. 7.4(a) for the simulated data and a schematic illustration of the simulation layout. In the simulation we use three microwave ports, located on the central qubit island, the qubit ring electrode, and on the coupling antenna of the lumped-element resonator on the upper chip. We deduce the effective coupling capacitance from the simulated admittance matrix. In a separate simulation, we find that the coupling strength is equal for various positions at the ring electrode of the concentric transmon qubit. This can be understood by the lumped-element nature of the concentric transmon qubit, see Ch. 5. By simulating the level repelling between the qubit and a bosonic mode resonator, we extract a maximum coupling strength $g_{i,\mathrm{max}}/2\pi = 1.0\,\mathrm{MHz}$ for $\Delta l = 0$ at a chip spacing of 35 µm, see Fig. 7.4(b).

The resistive elements for the bosonic bath resonators are formed by a gold-palladium AuPd film, which does not become superconducting at low temperatures. The internal quality factor $Q_{\mathrm{i},i} = Z_i/R$ of bosonic mode i can be calculated from the series resistance R of the normal metal piece.

On a silicon substrate we measure a room temperature square resistance $R_\square = 2.6\,\Omega$ for a AuPd film thickness of 85 nm. With a designed square resistance of 0.25 □ in the bosonic bath resonators, we obtain a target resistance of 0.51 Ω at base temperature, taking into account a decrease in the sheet resistance by 20% [Kle15].

Micrographs of fabricated bosonic baths are depicted in Fig. 7.3(c)-(f). The individual resonators are radially arranged around the centre of the concentric transmon on the lower chip, making use of its rotational symmetry. We either arrange the resonators in a spiral configuration such that adjacent ones are also neighbouring in frequency, Fig. 7.3(c), or with a respective maximum distance of resonators that are neighbouring in frequency, see Fig. 7.3(f). The motivation for distributing the single resonators is to minimize the parasitic coupling $C_{\mathrm{p},i}$. Figures 7.3(c)-(e) show a bosonic bath version with a resistive copper circuit element, which was not measured during this thesis. Measurement data with the spiral configuration is not shown here. Table 7.1 summarizes the relevant design parameters of the bosonic bath.

7.6.1 Circuit quantization

We can find the environment impedance $Z_\mathrm{b}(\omega)$ with the schematic circuit depicted in Fig. 7.3(b). Following Ref. [Lep+17] and evaluating Kirchhoff's current conservation law at the nodes above the single resonator impedances Z_i with potential difference V_i yields

$$\frac{V - V_i}{Z_{C_i}} = \frac{V_i - V_{i-1}}{Z_{p,i-1}} + \frac{V_i - V_{i+1}}{Z_{p,i}} + \frac{V_i}{Z_i} \tag{7.25}$$

for $i = 1..N$ with N the number of resonators. $Z_{\mathrm{p},i} = 1/i\omega C_{\mathrm{p},i}$ denotes the impedance of the parasitic capacitive coupling $C_{\mathrm{p},i}$ between node i and node $i+1$, and $Z_{C_i} = 1/\mathrm{i}\omega C_{\mathrm{c},i}$ is the impedance of the coupling capacitor with capacitance $C_{\mathrm{c},i}$. Diagonalization of Eq. (7.25) yields

$$Z_\mathrm{b}^{-1} = \frac{I}{V} = \frac{1}{V}\sum_i \frac{V - V_i}{Z_{C,i}}. \tag{7.26}$$

assuming a voltage drop V across the bath impedance [Lep+17].

7.6.2 Alternative approaches

During the course of this thesis, we investigated and evaluated different approaches for bosonic bath engineering. In the following, some efforts in alternative directions are outlined.

Fano resonance

A similar approach as the one described involves a single primary resonator of low internal quality factor Q_i, that couples to the qubit. Its frequency response is altered by introducing one or many secondary resonators that only couple to the primary resonator but not to the qubit and ideally do not couple mutually. By this, the initially Lorentzian frequency response of the primary resonator is reshaped in a controlled way, in order to recover the desired bath spectral density $J(\omega)$ within a defined frequency band [Hau15]. Numerical simulations were carried out within an undergraduate theses [Fri15], revealing promising results of the approach. However, microwave simulations and initial experiments [Pis15] showed that the frequency response is very sensitive on experimental uncertainties and crosstalk between single resonators. In addition, the spectral width of the primary resonator needs to be as large as the desired spectral width of the bath, which renders it extremely dissipative and therefore its coupling to the quantum circuit becomes very small.

By adjusting the couplings between qubit and primary resonator and between primary and secondary resonators, a dedicated shaping of the joint frequency response can be achieved by making use of Fano dynamics. In a system of two coupled oscillators with one of them being subject to a driving force, one can observe an asymmetric profile in the frequency response of the driven oscillator [MFK10].

Bath shaping via the introduction of an impedance mismatch

An elegant approach to engineer the spectral function of a bosonic bath is to use a 50 Ω transmission line and tailor its $J(\omega)$ by the introduction of controlled impedance mismatches ΔZ. Two such discontinuities can define a spacially restricted area of the line with a reduced number of modes.

A large enough ΔZ, yields an in-line resonator with a single fundamental mode. This allows one to tune between a very sharp frequency response (for a strong impedance mismatch) and the ohmic spectral function of a 50 Ω transmission line. Choosing asymmetric coupling $\Delta Z_1 \neq \Delta Z_2$, where $\Delta Z_{1,2}$ are the two impedance mismatches defining the resonating region, allows for an asymmetric shaping of the resonator's frequency response.

The coupling of transmission lines to a qubit and the frequency selection of the bath is however challenging. One approach can be to capacitively couple a half open transmission line to the qubit. The transmission line however would not have a tailored cut-off at small and large frequencies, respectively. In order to suppress the coupling between qubit and transmission line within a defined region, an additional microwave filter may be used, which renders the approach rather challenging.

7.7 Preliminary spin boson simulator

For the preliminary spin boson experiment, we couple a concentric transmon qubit to the bosonic bath including resistive AuPd elements in the configuration where frequency adjacent resonators are arranged at maximum distance, see Fig. 7.3(f). Figure 7.5(a) shows a micrograph of both samples mounted inside the sample box and a magnified view of the aligned chips is depicted in Fig. 7.5(b). Manual alignment was performed under the optical microscope.

The coupling between qubit and readout resonator is measured by spectroscopically recording their avoided crossing when tuning them on resonance, see Fig. 7.5(c). The fit yields a coupling strength of $g/2\pi = 63.5\,\mathrm{MHz}$.

Figure 7.5(d) shows the measured qubit decay rate Γ_1 with respect to the qubit transition frequency ϵ. As depicted by the grey area in Fig 7.5(d), we calculate a lower bound of the qubit decay rate induced by the single-mode Purcell effect of the readout resonator, based on its measured line width $\kappa = \kappa(\epsilon)$. At frequencies smaller than 7 GHz and larger than 7.8 GHz, the decay rate roughly follows the single-mode Purcell limitation.

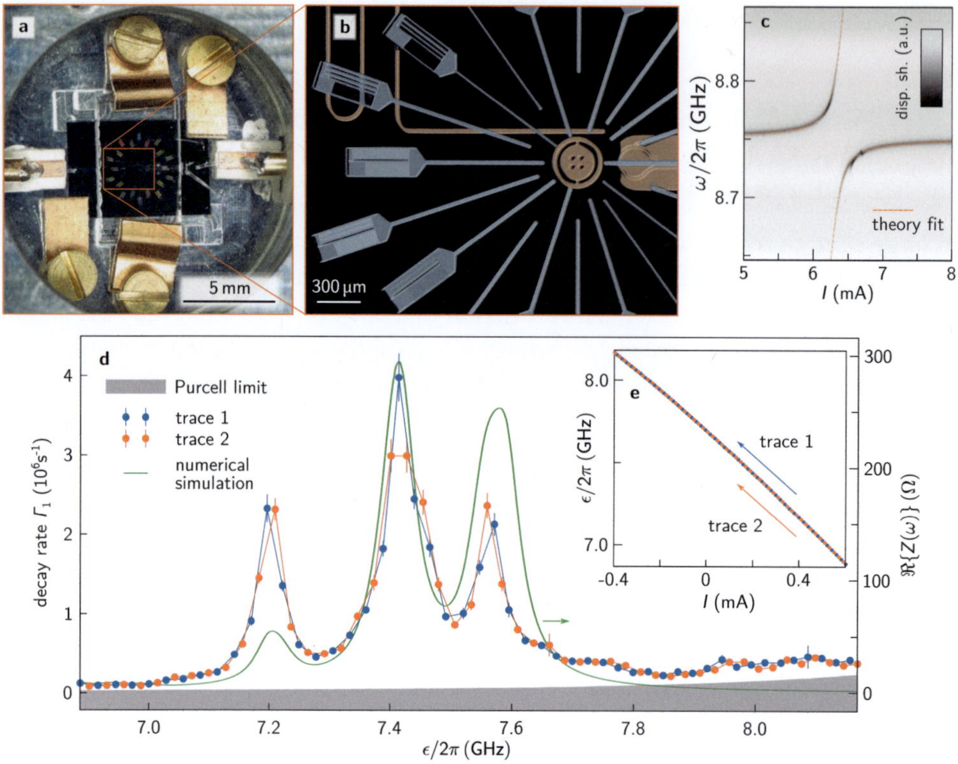

Figure 7.5: Spin boson experiment (a) Micrograph showing the qubit sample and the AuPd bosonic bath sample mounted in the sample box. (b) Closer view of the chip stack. The lower qubit sample is coloured in orange while the bosonic bath sample is coloured in blue. The bosonic bath is aligned in the centre symmetric projection of the concentric transmon qubit. (c) Avoided level crossing between qubit and readout resonator, yielding a coupling strength of $g/2\pi = 63.5\,\text{MHz}$. (d) Measured qubit decay rate $\Gamma_1 = 1/T_1$ dependent on the qubit transition frequency ϵ. In a band of $\sim 0.5\,\text{GHz}$ we observe a strong increase in the decay rate due to the presence of the bosonic bath. The three-peak structure qualitatively agrees with a numerical simulation of the quantized circuit and is attributed to the parasitic coupling of individual bosonic bath resonators. The grey area on the bottom of the figure shows the expected qubit decay rate only taking into account the single-mode Purcell effect of the readout resonator based on the coupling strength extracted in panel (c). Blue and orange points correspond to successive measurement traces as indicates in the inset (e), demonstrating that the measured signature is not due to temporal fluctuations in Γ_1. Error bars are scaled up by a factor of ten for better visibility. (e) Qubit dispersion versus the applied bias current I. Blue and orange data points again correspond to the consecutively measured traces.

In a frequency band of $\sim 0.5\,\text{GHz}$ we observe an increased decay rate induced by the bosonic bath. Apart from a region at 7.3 GHz, the decay rate is enhanced by more than one order of magnitude within the spectral location of the bosonic bath. The three-peak substructure is qualitatively reproduced by a numerical simulation of the quantized circuit. Numerical data of the real part of the bath impedance $\Re Z(\omega) \sim S(\omega)$ is shown as a green line in Fig. 7.5(d). The simulation takes into account the exact circuit design of the bosonic bath and includes accordingly varying mutual capacitive couplings $C_{\mathrm{p},i}$ between individual bosonic bath resonators. Deviations between measured data and the numerical simulation are attributed to fabrication tolerances and an imprecise alignment of the two chips.

From varying the spectral width in numerical simulations, we can observe a frequency bunching of the bosonic bath resonators due to their mutual coupling, leading to the formation of three peaks in the resulting spectral function. This indicates that the observed shape in the measured qubit decay rate originates from the response of many individual resonators that form the bosonic bath, rather than being induced by three individual resonators.

Blue and orange data points in Fig. 7.5(d) were acquired in two successive measurement traces. Since both traces reproduce the same substructure in Γ_1, we can exclude temporal fluctuations in measured data. A frequency offset in the bosonic bath response on the order of $\sim 400\,\mathrm{MHz}$ is attributed to an unconsidered capacitance coupling of the individual resonators to the sample box ground or a lithography bias.

Figure 7.5(e) shows the qubit transition frequency ϵ with respect to the dc bias current I applied to the flux bias coil. Data points correspond to fits of the qubit transition frequency, as measured during acquisition of the measurement sequence in Fig. 7.5(d).

For every data point depicted in Fig. 7.5(d), we record the qubit frequency ϵ, calibrate the π-pulse in a Rabi measurement and measure the inverse decay rate in a qubit lifetime measurement. As visible in Fig. 7.6(a), we use rectangular π-pulses of $\sim 50\,\mathrm{ns}$ duration. The window size Δt in the qubit lifetime measurements presented in Fig. 7.5(d) was $20\,\mu\mathrm{s}$, see Fig. 7.6(b). In a separate measurement, we validate the increased lifetime of the qubit state outside of the spectral band of the bosonic bath, see Fig. 7.6(c).

In a separate experiment, we probed purely superconducting versions of fabricated bosonic baths by measuring microwave transmission through a line that couples to the bosonic bath. We find the response of ~ 10 individual resonators within a spectral region of 0.5 GHz, which supports the above indicated multi-mode response. Via the relation between the transmission matrix element S_{21}^{n} and the impedance of the scatterer, see Eq. (3.23) in Sec. 3.4.3, we can map the acquired transmission data to the spectral function $S(\omega)$ of the environment.

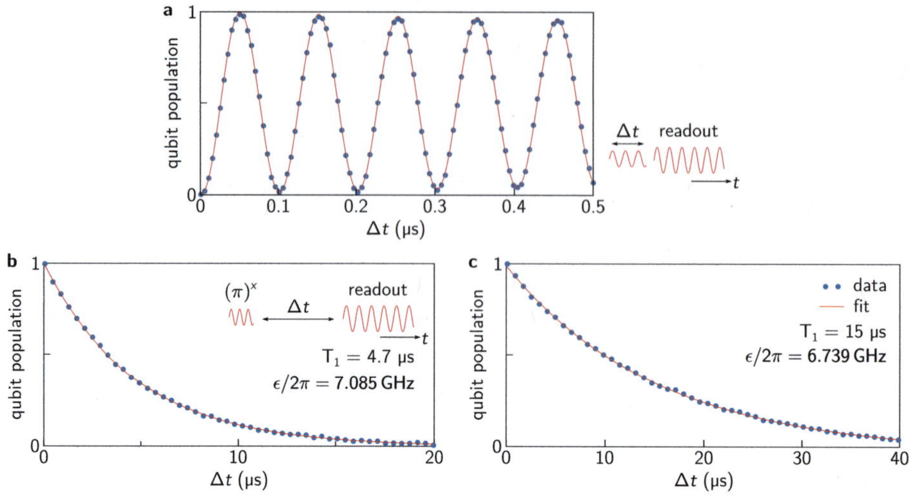

Figure 7.6: Rabi and decay time measurements during the spin boson experiment (a) Illustrative Rabi driving of the qubit during the measurement depicted in Fig. 7.5. The qubit was located at 7.085 GHz. We apply pulses with a rectangular envelope, and use π-pulses of $\sim$ 50 ns duration. (b) Respective lifetime measurement yielding $T_1 = 4.7\,\mu s$ and $\Gamma_1 = 0.21 \times 10^6\,s^{-1}$. The window size for the lifetime measurement in Fig. 7.5 was 20 μs. (c) The maximum lifetime for the qubit is measured at a frequency below the spectral location of the bosonic bath. With the qubit biased to 6.739 GHz, we find $T_1 = 15\,\mu s$ and $\Gamma_1 = 0.065 \times 10^6\,s^{-1}$.

7.8 Conclusion and Outlook

We have presented an experimental approach that allows for the circuit implementation of the spin boson model. We engineered a dissipative bosonic environment using a set of 20 lumped-element resonators. By equipping the individual bosonic modes with resistive circuit elements, their line width is increased, such that they overlap and form a bosonic bath of continuous spectral function.

In our experiment, we fabricated the bosonic bath and the qubit on two separate chips. They are attached on top of each other at a spacing of 35 μm in a specifically designed sample box. The upper chip, holding the bosonic bath, is flipped upside down and therefore facing the qubit sample on the bottom. The approach makes use of the rotational symmetry of the concentric transmon qubit, which allows to couple many individual resonators.

In the preliminary spin boson experiment, we observe a strong increase in the qubit decay rate induced by the engineered bosonic bath within a spectral band of 0.5 GHz width. We observed a pronounced substructure that is caused by a frequency bunching of individual resonators, in qualitative agreement with a numerical simulation of the quantized circuit. While a frequency bunching tendency has been found theoretically in disordered systems [KMS16], we attribute the effect to the mutual couplings in the bosonic bath circuit. The simulations therewith indicate the desired many-mode response, validating our approach.

The presented experiment is a starting point to tailor specific shapes of spectral functions in successive experiments. We are confident that the described frequency bunching can be avoided by taking into account the parasitic mode couplings in circuit design.

There is an ongoing discussion whether the spectral function of the bosonic bath must be a smooth continuum of modes in order to recover the spin boson model. A bosonic bath formed by distinct sharp modes with a separation comparable to the coupling strength of the single modes to the qubit may equally map to the spin boson model in the thermodynamic limit [Roc17]. While this being a pending question, the modular approach presented here can also be applied for bosonic baths of high internal quality. This furthermore allows us to make use of the pre-characterisation scheme using a separate probe sample.

The spin boson model exhibits its critical phenomena in a regime where the effective Kondo parameter of the bosonic environment is close to unity [Lep+17]. This can be achieved by creating ultra-strong coupling in the spin boson model, corresponding to a regime where the energy decay rate of the two-level system is comparable to its effective transition frequency [For+16a; Lep+17]. For this purpose, we are currently exploring the feasibility of applying a similar driving scheme as presented in Ch. 6.

8 Conclusion

It was the goal of this thesis to apply few basic elements from the toolbox of superconducting quantum circuits and demonstrate that useful manipulations in the spirit of analog quantum simulations can be performed. The presented proof-of-principle experiments demonstrate the large potential of superconducting circuits for building quantum hardware that is capable of efficiently simulating physical phenomena.

In the course of this thesis we have started with designing a planar tunable qubit that we call concentric transmon [Bra+16]. It stands out by forming a robust building block for quantum information experiments with a straightforward fabrication process and is competitive in coherence with conventional transmon geometries. We could reproducibly measure quantum excitation lifetimes of up to 15 µs in our fabricated devices. The qubit lifetimes are limited by the Purcell effect and loss induced by interface and surface defects. As compared to conventional transmon designs, fields are more confined in the concentric layout, which reduces coupling to spurious modes in the electromagnetic environment of the sample. The versatility of the concentric transmon in particular for quantum simulation experiments becomes most evident given its potential for the implementation of a direct site-selective $\hat{Z}$ coupling between adjacent qubits. The presence of a longitudinal $\hat{Z}$ coupling is under current experimental investigation [Ste16].

Quantum simulations that were performed in this thesis are based on the quantum Rabi model and the spin boson model that generically describe the fundamental light-matter interaction on a quantum level. These were the models of choice because they are ubiquitous in nature and provide non-trivial physics that can be studied within the framework of quantum simulation at a small experimental expense. They require only few technological and hardware resources that are available in the toolbox of superconducting circuits, namely a single qubit and several harmonic resonator modes, as in our approach.

The main achievement of this thesis is the analog quantum simulation of the quantum Rabi model in the ultra-strong coupling regime [Bra+17]. While this extreme coupling regime is very challenging to access geometrically by sample design, we achieved ultra-strong coupling (USC) in an effective quantum Rabi model with decreased sub-system energies. The applied simulation scheme is based on the application of two microwave Rabi drives that couple to the qubit. In the effective frame, the synthesized model is operated at a relative coupling regime of up to $g_{\mathrm{eff}}/\omega_{\mathrm{eff}} \sim 0.6$, approaching the deep strong coupling regime. As the hallmark signature of USC in the quantum Rabi model we have observed quantum state collapses and revivals in the qubit dynamics. The non-conservation of the total excitation number in the circuit, imposed by the counter-rotating terms in the Hamiltonian, has been indirectly observed. Limitations of the simulation scheme became apparent for a finite qubit energy term in the effective Hamiltonian. However, we could obtain experimental evidence of the desired signatures in our quantum simulation. The accessible coupling regime was not fundamentally limited by the applied scheme, but rather by the finite coherence in our circuit. While the system dynamics in the effective frame is slowed down, the laboratory frame dissipation rates are maintained in the synthesized frame. This comes at the expense of a small cooperativity measure which is an ultimate limitation of the simulation scheme. The decelerated system dynamics however allowed us to observe quantum revivals at a rate that is compatible with our experimental resolution. Revivals in the laboratory frame USC quantum Rabi model would occur on a time scale of $2\pi\omega^{-1} \sim 0.2\,\mathrm{ns}$ and are therewith impossible to resolve with conventional experimental equipment.

In the final experimental chapter of this thesis I presented first steps in extending the quantum Rabi model by substituting the single quantized oscillator mode with an engineered bosonic bath, recovering the spin boson model. In our experimental approach, we coupled a concentric transmon qubit to a set of 20 resonators that form a dissipative environment. The individual series lumped-element resonators are equipped with a resistive element that increases their spectral width such that they overlap and form a continuous spectral function. A shaping of the spectral function is achieved by varying the coupling between individual resonators to the qubit. The qubit and the bosonic bath were fabricated on two separate chips in a modular flip-chip approach. They are attached on top of each other by making use of a specifically designed sample box, while both chips are facing each other to enhance their mutual coupling. The bosonic bath is therewith physically separated from the qubit and its readout and control circuitry, in contrast to previous experiments [Hae+15; For+16a]. We presented experimental evidence of a strong increase in the qubit decay rate due to an engineered bosonic environment. The shape of the spectral function in the initial experiment was dominated by a three-peak structure, in qualitative agreement with a numerical simulation of the quantized circuit. Simulations have shown that the substructure is caused by a frequency bunching of bosonic bath resonators, thereby indicating the desired many-mode response. The presented experiment is a starting point to tailor specific shapes of spectral functions by taking into account the parasitic mode couplings in circuit design.

In conclusion, we have succeeded to perform several proof-of-principle experiments simulating interesting quantum models of relevance in nature, and demonstrated the potential of superconducting circuits for the field of quantum simulation.

Outlook

An experimental implementation of the spin boson model at arbitrary coupling and with a tailored spectral function would be a rich platform to study interesting phenomena and dynamics of the model. A similar drive scheme as applied to create USC in the quantum Rabi model can be a route to achieve coupling conditions with a Kondo parameter approaching unity also in the spin boson model [Lep+17]. By again creating an effective model in a rotating frame with decreased subsystem energies, the zero-frequency properties of the spectral function can additionally be well controlled. This corresponds to choosing the effective system temperature [Lep+17].

In general, the spin boson model exhibits a multitude of interesting phenomena [OIL13]. At sufficiently large coupling to the bosonic environment, the spin undergoes a quantum phase transition from a de-localized to a localized phase, which manifests itself in new ground state properties of the two-level system, similar to the dynamics we observed in the quantum Rabi model at USC. The corresponding dynamics is accessible in experiment and appears as a finite expectation value of the well population function [FVN10; Leg+87; Lep+17].

For an ohmic spectral density, the spin boson model can be exactly mapped to the anisotropic Kondo model via bosonization, which describes a procedure valid in one dimension where a set of non-interacting bosons is transformed to a system of interacting fermions in the spirit of Luttinger liquid theory [BEL70; Le 12]. This opens up the possibility to experimentally access and simulate a range of impurity models relevant in condensed matter physics, such as the Kondo problem itself, the Anderson impurity model or the one-dimensional Ising model [OIL13], as well as the paradigmatic Mott-insulator transition in the Hubbard model [HTK12].

The main impediment to advance an implementation of the universal quantum computer with superconducting circuits is energy dissipation and decoherence due to an uncontrolled bosonic environment [OIL13]. A spin boson simulator therefore may provide a generic tool to study and address these limitations.

It remains as one of the open questions to the field how to properly quantify or benchmark the performance of a quantum simulation in general. Assessing the result of a quantum computation that cannot be reproduced and checked by classical means is evidently difficult. Even the benchmarking of analog quantum simulators that are still within reach of the capabilities of classical computers is not entirely obvious. Stating the standard deviation between the experimental result and a numerical calculation may not do justice to a quantum simulator suffering from experimental errors and noise but is capable of providing qualitative signatures of the investigated quantum system.

While this thesis focussed on the approach of analog quantum simulation, it is an ongoing debate how to first reach quantum supremacy in a useful computation. The digital simulation approach may be considered to be more exact in the sense that quantum hardware is exploited to provide fault-tolerant logical quantum elements. Error monitoring and correction however comes at the cost of requiring massive hardware resources that exceed the capabilities of present technologies. Given the finite coherence in present superconducting circuits and other quantum hardware platforms of similar controllability, the number of gates within one computational step is strongly limited in the digital approach. The consequence is a very restricted level of error-correction and a coarse Trotter decomposition that comes at the expense of poor accuracy in mapping a complex quantum evolution.

The analog approach equally suffers from the finite coherence in an engineered circuit. One can however conceive a scenario where a noisy environment that leads to decoherence of physical qubits may be considered as part of the simulator. This appears to be particularly appealing when emulating complex mechanisms of protein complexes or molecules in quantum chemistry that are naturally coupling to a lossy environment.

It is my conviction that the universal quantum computer will be a truly digital device while meaningful computations with quantum hardware in the near future will be based on analog simulation schemes. This thesis presents a small contribution in this direction.

Bibliography

[AR16] C. H. Alderete and B. M. Rodríguez-Lara: *Cross-cavity quantum Rabi model*. J. Phys. A: Math. Theor. **49**.41 (2016), p. 414001. URL: `http://stacks.iop.org/1751-8121/49/i=41/a=414001` (cit. on p. 74).

[ABV07] F. B. Anders, R. Bulla, and M. Vojta: *Equilibrium and Nonequilibrium Dynamics of the Sub-Ohmic Spin-Boson Model*. Phys. Rev. Lett. **98** (2007), p. 210402. DOI: `10.1103/PhysRevLett.98.210402` (cit. on pp. 5, 99).

[AS06] F. B. Anders and A. Schiller: *Spin precession and real-time dynamics in the Kondo model: Time-dependent numerical renormalization-group study*. Phys. Rev. B **74** (2006), p. 245113. DOI: `10.1103/PhysRevB.74.245113` (cit. on p. 99).

[Ann11] J. F. Annett: *Supraleitung, Suprafluidität und Kondensate*. Oldenbourg Verlag, 2011 (cit. on p. 15).

[AM76] N. W. Ashcroft and N. D. Mermin: *Solid State Physics*. Saunders College Philadelphia, 1976 (cit. on p. 13).

[AN10] S. Ashhab and F. Nori: *Qubit-oscillator systems in the ultrastrong-coupling regime and their potential for preparing nonclassical states*. Phys. Rev. A **81**.4 (2010), p. 042311. DOI: `10.1103/physreva.81.042311` (cit. on p. 74).

[Ast64] A. V. Astin: *Handbook of Mathematical Functions with Formulas, Graphs, and Mathematical Tables*. Ed. by M. Abramowitz and I. A. Stegun. National Bureau of Standards, Washington, D.C., 1964 (cit. on p. 22).

[Bab+15] R. Babbush, J. McClean, D. Wecker, A. Aspuru-Guzik, and N. Wiebe: *Chemical basis of Trotter-Suzuki errors in quantum chemistry simulation*. Phys. Rev. A **91** (2015), p. 022311. DOI: `10.1103/PhysRevA.91.022311` (cit. on pp. 2, 4).

[Bal+12] D. Ballester, G. Romero, J. J. García-Ripoll, F. Deppe, and E. Solano: *Quantum Simulation of the Ultrastrong-Coupling Dynamics in Circuit Quantum Electrodynamics*. Phys. Rev. X **2**.2 (2012), p. 021007. DOI: `10.1103/physrevx.2.021007` (cit. on pp. 5, 74, 75, 77, 98).

[BCS57] J. Bardeen, L. N. Cooper, and J. R. Schrieffer: *Theory of Superconductivity*. Phys. Rev. **108**.5 (1957). DOI: `10.1103/PhysRev.108.1175` (cit. on p. 13).

[Bar+13] R. Barends, J. Kelly, A. Megrant, D. Sank, E. Jeffrey, Y. Chen, Y. Yin, B. Chiaro, J. Mutus, C. Neill, P. O'Malley, P. Roushan, J. Wenner, T. White, A. N. Cleland, and J. M. Martinis: *Coherent Josephson Qubit Suitable for Scalable Quantum Integrated Circuits*. Phys. Rev. Lett. **111**.8 (2013), p. 080502. DOI: 10.1103/physrevlett.111.080502 (cit. on pp. 65, 69).

[Bar+14] R. Barends, J. Kelly, A. Megrant, A. Veitia, D. Sank, E. Jeffrey, T. C. White, J. Mutus, A. G. Fowler, B. Campbell, Y. Chen, B. Chiaro, A. Dunsworth, C. Neill, P. O'Malley, P. Roushan, A. Vainsencher, J. Wenner, A. N. Korotkov, A. N. Cleland, and J. M. Martinis: *Superconducting quantum circuits at the surface code threshold for fault tolerance*. Nature **508**.7497 (2014), pp. 500–503. DOI: 10.1038/nature13171 (cit. on pp. 4, 53).

[Bar+15] R. Barends, L. Lamata, J. Kelly, L. García-Álvarez, A. G. Fowler, A. Megrant, E. Jeffrey, T. C. White, D. Sank, J. Y. Mutus, B. Campbell, Y. Chen, Z. Chen, B. Chiaro, A. Dunsworth, I.-C. Hoi, C. Neill, P. J. J. O'Malley, C. Quintana, P. Roushan, A. Vainsencher, J. Wenner, E. Solano, and J. M. Martinis: *Digital quantum simulation of fermionic models with a superconducting circuit*. Nat. Commun. **6** (2015), p. 7654. DOI: 10.1038/ncomms8654 (cit. on p. 4).

[BGB11] F. Beaudoin, J. M. Gambetta, and A. Blais: *Dissipation and ultrastrong coupling in circuit QED*. Phys. Rev. A **84**.4 (2011), p. 043832. DOI: 10.1103/physreva.84.043832 (cit. on p. 74).

[Bla+04] A. Blais, R.-S. Huang, A. Wallraff, S. M. Girvin, and R. J. Schoelkopf: *Cavity quantum electrodynamics for superconducting electrical circuits: An architecture for quantum computation*. Phys. Rev. A **69**.6 (2004), p. 062320. DOI: 10.1103/physreva.69.062320 (cit. on pp. 23, 25, 26, 78).

[BEL70] M. Blume, V. J. Emery, and A. Luther: *Spin-Boson Systems: One-Dimensional Equivalents and the Kondo Problem*. Phys. Rev. Lett. **25** (1970), pp. 450–453. DOI: 10.1103/PhysRevLett.25.450 (cit. on p. 119).

[Boi+18] S. Boixo, S. V. Isakov, V. N. Smelyanskiy, R. Babbush, N. Ding, Z. Jiang, M. J. Bremner, J. M. Martinis, and H. Neven: *Characterizing quantum supremacy in near-term devices*. Nat. Phys. (2018). DOI: 10.1038/s41567-018-0124-x (cit. on p. 2).

[Bou+98] V. Bouchiat, D. Vion, P. Joyez, D. Esteve, and M. H. Devoret: *Quantum coherence with a single Cooper pair*. Phys. Scr. **1998**.T76 (1998), p. 165. URL: http://stacks.iop.org/1402-4896/1998/i=T76/a=024 (cit. on p. 20).

[Bra11] D. Braak: *Integrability of the Rabi Model*. Phys. Rev. Lett. **107**.10 (2011), p. 100401. DOI: 10.1103/physrevlett.107.100401 (cit. on pp. 24, 73).

[Bra+15] J. Braumüller, J. Cramer, S. Schlör, H. Rotzinger, L. Radtke, A. Lukashenko, P. Yang, S. T. Skacel, S. Probst, M. Marthaler, L. Guo, A. V. Ustinov, and M. Weides: *Multiphoton dressing of an anharmonic superconducting many-level quantum circuit*. Phys. Rev. B **91**.5 (2015), p. 054523. DOI: 10.1103/physrevb.91.054523 (cit. on pp. 22, 23, 27, 62, 82).

[Bra+17] J. Braumüller, M. Marthaler, A. Schneider, A. Stehli, H. Rotzinger, M. Weides, and A. V. Ustinov: *Analog quantum simulation of the Rabi model in the ultra-strong coupling regime*. Nat. Commun. **8**.1 (2017), p. 779. DOI: 10.1038/s41467-017-00894-w (cit. on pp. 5, 73, 77, 93, 117).

[Bra+16] J. Braumüller, M. Sandberg, M. R. Vissers, A. Schneider, S. Schlör, L. Grünhaupt, H. Rotzinger, M. Marthaler, A. Lukashenko, A. Dieter, A. V. Ustinov, M. Weides, and D. P. Pappas: *Concentric transmon qubit featuring fast tunability and an anisotropic magnetic dipole moment*. Appl. Phys. Lett. **108**.3 (2016), p. 032601. DOI: 10.1063/1.4940230 (cit. on pp. 5, 53, 67–69, 78, 117).

[Bra13] J. Braumüller: *Development of a tunable transmon qubit in microstrip geometry*. MA thesis. Karlsruher Institut für Technologie, 2013 (cit. on p. 22).

[BCR11] C. Brif, R. Chakrabarti, and H. Rabitz: *Advances in Chemical Physics*. Ed. by S. A. Rice and A. R. Dinner. Vol. 148. Wiley, 2011. Chap. Control of quantum phenomena. DOI: 10.1002/9781118158715.ch1 (cit. on p. 9).

[BN09] I. Buluta and F. Nori: *Quantum Simulators*. Science **326**.5949 (2009), pp. 108–111. DOI: 10.1126/science.1177838 (cit. on p. 3).

[Byr+08] T. Byrnes, N. Kim, K. Kusudo, and Y. Yamamoto: *Quantum simulation of Fermi-Hubbard models in semiconductor quantum-dot arrays*. Phys. Rev. B **78**.7 (2008). DOI: 10.1103/physrevb.78.075320 (cit. on p. 4).

[CL83] A. Caldeira and A. Leggett: *Quantum tunnelling in a dissipative system*. Ann. Phys. **149**.2 (1983), pp. 374–456. DOI: 10.1016/0003-4916(83)90202-6 (cit. on p. 100).

[Cas+10] J. Casanova, G. Romero, I. Lizuain, J. J. García-Ripoll, and E. Solano: *Deep Strong Coupling Regime of the Jaynes-Cummings Model*. Phys. Rev. Lett. **105**.26 (2010), p. 263603. DOI: 10.1103/physrevlett.105.263603 (cit. on pp. 5, 73, 74).

[Cha+13] J. B. Chang, M. R. Vissers, A. D. Córcoles, M. Sandberg, J. Gao, D. W. Abraham, J. M. Chow, J. M. Gambetta, M. Beth Rothwell, G. A. Keefe, M. Steffen, and D. P. Pappas: *Improved superconducting qubit coherence using titanium nitride*. Appl. Phys. Lett. **103**.1 (2013), p. 012602. DOI: 10.1063/1.4813269 (cit. on p. 68).

[Cho+12] J. M. Chow, J. M. Gambetta, A. D. Córcoles, S. T. Merkel, J. A. Smolin, C. Rigetti, S. Poletto, G. A. Keefe, M. B. Rothwell, J. R. Rozen, M. B. Ketchen, and M. Steffen: *Universal Quantum Gate Set Approaching Fault-Tolerant Thresholds with Superconducting Qubits*. Phys. Rev. Lett. **109**.6 (2012), p. 060501. DOI: 10.1103/physrevlett.109.060501 (cit. on p. 69).

[CZ03] J. I. Cirac and P. Zoller: *How to Manipulate Cold Atoms*. Science **301**.5630 (2003), pp. 176–177. DOI: 10.1126/science.1085130 (cit. on p. 2).

[CW08] J. Clarke and F. K. Wilhelm: *Superconducting quantum bits*. Nature **453** (2008). DOI: 10.1038/nature07128 (cit. on pp. 4, 19, 20).

[Col+10] J. H. Cole, C. Müller, P. Bushev, G. J. Grabovskij, J. Lisenfeld, A. Lukashenko, A. V. Ustinov, and A. Shnirman: *Quantitative evaluation of defect-models in superconducting phase qubits*. Appl. Phys. Lett. **97**.25 (2010), p. 252501. DOI: 10.1063/1.3529457 (cit. on p. 65).

[CL04] N. Craig and T. Lester: *Hitchhiker's guide to the dilution refrigerator*. Harward University, 2004 (cit. on p. 46).

[Deu85] D. Deutsch: *Quantum Theory, the Church-Turing Principle and the Universal Quantum Computer*. *Proc. R. Soc. Lond. A*. Vol. 400. 1818. The Royal Society, 1985, pp. 97–117. DOI: 10.1098/rspa.1985.0070 (cit. on p. 1).

[Dev97] M. H. Devoret: *Quantum fluctuations. Les Houches, Session LXIII*. Ed. by E. G. S. Reynaud and J. Zinn-Justin. Elsevier Science B. V., 1997. Chap. Quantum fluctuations in electrical circuits, pp. 351–386. URL: http://qulab.eng.yale.edu/documents/reprints/Houches_fluctuations.pdf (cit. on pp. 17, 18).

[DS13] M. H. Devoret and R. J. Schoelkopf: *Superconducting ciruits for quantum information: an outlook*. Science **339**.6124 (2013), pp. 1169–1174. DOI: 10.1126/science.1231930 (cit. on p. 4).

[Dol77] G. J. Dolan: *Offset masks for lift-off photoprocessing*. Appl. Phys. Lett. **31**.5 (1977), pp. 337–339. DOI: 10.1063/1.89690 (cit. on p. 44).

[ENS80] J. H. Eberly, N. B. Narozhny, and J. J. Sanchez-Mondragon: *Periodic Spontaneous Collapse and Revival in a Simple Quantum Model*. Phys. Rev. Lett. **44**.20 (1980), pp. 1323–1326. DOI: 10.1103/physrevlett.44.1323 (cit. on p. 85).

[Eng+07] G. S. Engel, T. R. Calhoun, E. L. Read, T.-K. Ahn, T. Mancal, Y.-C. Cheng, R. E. Blankenship, and G. R. Fleming: *Evidence for wavelike energy transfer through quantum coherence in photosynthetic systems*. Nature **446**.7137 (2007), pp. 782–786. URL: http://dx.doi.org/10.1038/nature05678 (cit. on p. 2).

[FI15] L. Faoro and L. B. Ioffe: *Interacting tunneling model for two-level systems in amorphous materials and its predictions for their dephasing and noise in superconducting microresonators*. Phys. Rev. B **91**.1 (2015), p. 014201. DOI: 10.1103/physrevb.91.014201 (cit. on p. 65).

[Fey82] R. P. Feynman: *Simulating physics with computers*. Int. J. Theor. Phys. **21**.6-7 (1982), pp. 467–488. DOI: 10.1007/BF02650179 (cit. on pp. 1–3).

[FG96] G. A. Finney and J. Gea-Banacloche: *Quantum suppression of chaos in the spin-boson model*. Phys. Rev. E **54** (1996), pp. 1449–1456. DOI: 10.1103/PhysRevE.54.1449 (cit. on p. 99).

[FVN10] S. Florens, D. Venturelli, and R. Narayanan: *Quantum Phase Transition in the Spin Boson Model. Quantum Quenching, Annealing and Computation*. Ed. by A. K. Chandra, A. Das, and B. K. Chakrabarti. Springer Berlin Heidelberg, 2010. Chap. 6, pp. 145–162. DOI: 10.1007/978-3-642-11470-0_6 (cit. on pp. 100, 101, 119).

[For+16a] P. Forn-Díaz, J. García-Ripoll, B. Peropadre, J.-L. Orgiazzi, M. Yurtalan, R. Belyansky, C. Wilson, and A. Lupascu: *Ultrastrong coupling of a single artificial atom to an electromagnetic continuum in the nonperturbative regime*. Nat. Phys. **13**.1 (2016), pp. 39–43. DOI: 10.1038/nphys3905 (cit. on pp. 5, 115, 118).

[For+10] P. Forn-Díaz, J. Lisenfeld, D. Marcos, J. J. García-Ripoll, E. Solano, C. J. P. M. Harmans, and J. E. Mooij: *Observation of the Bloch-Siegert Shift in a Qubit-Oscillator System in the Ultrastrong Coupling Regime*. Phys. Rev. Lett. **105** (2010), p. 237001. DOI: 10.1103/PhysRevLett.105.237001 (cit. on p. 74).

[For+16b] P. Forn-Díaz, G. Romero, C. J. P. M. Harmans, E. Solano, and J. E. Mooij: *Broken selection rule in the quantum Rabi model*. Sci. Rep. **6**.1 (2016), p. 26720. DOI: 10.1038/srep26720 (cit. on p. 74).

[Fow+12] A. G. Fowler, M. Mariantoni, J. M. Martinis, and A. N. Cleland: *Surface codes: Towards practical large-scale quantum computation*. Phys. Rev. A **86** (2012), p. 032324. DOI: 10.1103/PhysRevA.86.032324 (cit. on p. 2).

[Fri+08] A. Friedenauer, H. Schmitz, J. T. Glueckert, D. Porras, and T. Schaetz: *Simulating a quantum magnet with trapped ions*. Nat. Phys. **4**.10 (2008), pp. 757–761. DOI: 10.1038/nphys1032 (cit. on p. 4).

[Fri15] L. Fritz: *Modellierung eines Bosonischen Bades*. Bachelor thesis. (2015) (cit. on p. 110).

[Gao08] J. Gao: *The physics of superconducting microwave resonators*. PhD thesis. California Institute of Technology, 2008 (cit. on pp. 37, 38).

[GAN14] I. M. Georgescu, S. Ashhab, and F. Nori: *Quantum simulation*. Rev. Mod. Phys. **86**.1 (2014), pp. 153–185. DOI: 10.1103/revmodphys.86.153 (cit. on pp. 1, 3, 4, 54).

[Ger+10] R. Gerritsma, G. Kirchmair, F. Zähringer, E. Solano, R. Blatt, and C. F. Roos: *Quantum simulation of the Dirac equation*. Nature **463**.7277 (2010), pp. 68–71. DOI: 10.1038/nature08688 (cit. on pp. 4, 74).

[Gil+08] M. Gilowski, T. Wendrich, T. Müller, C. Jentsch, W. Ertmer, E. M. Rasel, and W. P. Schleich: *Gauss Sum Factorization with Cold Atoms*. Phys. Rev. Lett. **100** (2008), p. 030201. DOI: 10.1103/PhysRevLett.100.030201 (cit. on p. 2).

[GL50] V. L. Ginzburg and L. D. Landau: *To the Theory of Superconductivity*. Zh. Eksp. Teor. Fiz **20**.1064 (1950) (cit. on p. 13).

[Göp+08] M. Göppl, A. Fragner, M. Baur, R. Bianchetti, S. Filipp, J. M. Fink, P. J. Leek, G. Puebla, L. Steffen, and A. Wallraff: *Coplanar waveguide resonators for circuit quantum electrodynamics*. J. Appl. Phys. **104**.11 (2008), p. 113904. DOI: 10.1063/1.3010859 (cit. on p. 37).

[Gra13] G. J. Grabovskij: *Investigation of coherent microscopic defects inside the tunneling barrier of a Josephson junction*. PhD thesis. Karlsruhe Institute of Technology (KIT), Physikaisches Institut, 2013 (cit. on p. 10).

[Gre+13] D. Greif, T. Uehlinger, G. Jotzu, L. Tarruell, and T. Esslinger: *Short-Range Quantum Magnetism of Ultracold Fermions in an Optical Lattice*. Science **340**.6138 (2013), pp. 1307–1310. DOI: 10.1126/science.1236362 (cit. on p. 4).

[Gre+02] M. Greiner, O. Mandel, T. Esslinger, T. W. Hänsch, and I. Bloch: *Quantum phase transition from a superfluid to a Mott insulator in a gas of ultracold atoms*. Nature **415**.6867 (2002), pp. 39–44. DOI: 10.1038/415039a (cit. on p. 4).

[Gul+11] E. Gull, A. J. Millis, A. I. Lichtenstein, A. N. Rubtsov, M. Troyer, and P. Werner: *Continuous-time Monte Carlo methods for quantum impurity models*. Rev. Mod. Phys. **83** (2011), pp. 349–404. DOI: 10.1103/RevModPhys.83.349 (cit. on p. 99).

[Hae+15] M. Haeberlein, F. Deppe, A. Kurcz, J. Goetz, A. Baust, P. Eder, K. Fedorov, M. Fischer, E. P. Menzel, M. J. Schwarz, F. Wulschner, E. Xie, L. Zhong, E. Solano, A. Marx, J.-J. García-Ripoll, and R. Gross: *Spin-boson model with an engineered reservoir in circuit quantum electrodynamics*. Preprint at `https://arxiv.org/abs/arXiv1506.09114`. (2015) (cit. on pp. 5, 118).

[Hah50] E. L. Hahn: *Spin Echoes*. Phys. Rev. **80** (1950), pp. 580–594. DOI: `10.1103/PhysRev.80.580` (cit. on p. 10).

[HR06] S. Haroche and J.-M. Raimond: *Exploring the Quantum - Atoms, Cavities, and Photons*. Oxford University Press, 2006 (cit. on pp. 9, 24, 85).

[Hau15] M. Hauck: *Emulation des Spin-Boson Modells in supraleitenden Schaltkreisen*. MA thesis. Karlsruhe Institute of Technology, 2015 (cit. on pp. 102, 110).

[Hen+17] T. Hensgens, T. Fujita, L. Janssen, X. Li, C. J. Van Diepen, C. Reichl, W. Wegscheider, S. Das Sarma, and L. M. K. Vandersypen: *Quantum simulation of a Fermi-Hubbard model using a semiconductor quantum dot array*. Nature **548**.7665 (2017), pp. 70–73. DOI: `10.1038/nature23022` (cit. on p. 4).

[Hou+08] A. A. Houck, J. A. Schreier, B. R. Johnson, J. M. Chow, J. Koch, J. M. Gambetta, D. I. Schuster, L. Frunzio, M. H. Devoret, S. M. Girvin, and R. J. Schoelkopf: *Controlling the Spontaneous Emission of a Superconducting Transmon Qubit*. Phys. Rev. Lett. **101**.8 (2008), p. 080502. DOI: `10.1103/physrevlett.101.080502` (cit. on pp. 62, 63).

[HTK12] A. A. Houck, H. E. Türeci, and J. Koch: *On-chip quantum simulation with superconducting circuits*. Nat. Phys. **8**.4 (2012), pp. 292–299. DOI: `10.1038/nphys2251` (cit. on pp. 4, 119).

[HP16] M.-J. Hwang and M. B. Plenio: *Quantum Phase Transition in the Finite Jaynes-Cummings Lattice Systems*. Phys. Rev. Lett. **117** (2016), p. 123602. DOI: `10.1103/PhysRevLett.117.123602` (cit. on p. 74).

[HPP15] M.-J. Hwang, R. Puebla, and M. B. Plenio: *Quantum Phase Transition and Universal Dynamics in the Rabi Model*. Phys. Rev. Lett. **115**.18 (2015), p. 180404. DOI: `10.1103/physrevlett.115.180404` (cit. on p. 74).

[Jaa+16] T. Jaako, Z.-L. Xiang, J. J. Garcia-Ripoll, and P. Rabl: *Ultrastrong-coupling phenomena beyond the Dicke model*. Phys. Rev. A **94** (2016), p. 033850. DOI: `10.1103/PhysRevA.94.033850` (cit. on p. 5).

[Jac62] J. Jackson: *Classical Electrodynamics*. New York: John Wiley & Sons, Inc., 1962 (cit. on pp. 55, 69).

[JC63] E. Jaynes and F. Cummings: *Comparison of quantum and semiclassical radiation theories with application to the beam maser*. Proc. IEEE **51**.1 (1963), pp. 89–109. DOI: `10.1109/proc.1963.1664` (cit. on pp. 24, 73).

[JNN12] J. Johansson, P. Nation, and F. Nori: *QuTiP: An open-source Python framework for the dynamics of open quantum systems*. Comp. Phys. Comm. **183**.8 (2012), pp. 1760–1772. DOI: `10.1016/j.cpc.2012.02.021` (cit. on p. 82).

[JNN13] J. Johansson, P. Nation, and F. Nori: *QuTiP 2: A Python framework for the dynamics of open quantum systems*. Comp. Phys. Comm. **184**.4 (2013), pp. 1234–1240. DOI: `10.1016/j.cpc.2012.11.019` (cit. on p. 82).

[Jos62] B. D. Josephson: *Possible new effects in superconductive tunnelling*. Phys. Lett. 1 **1**.7 (1962), pp. 251–253. DOI: `10.1016/0031-9163(62)91369-0` (cit. on p. 18).

[Kam11] H. Kammerlingh-Onnes: *The resistance of pure mercury at helium temperatures*. Commun. Phys. Lab. Univ. Leiden **12** (1911), p. 120 (cit. on p. 13).

[Kha+12] M. S. Khalil, M. J. A. Stoutimore, F. C. Wellstood, and K. D. Osborn: *An analysis method for asymmetric resonator transmission applied to superconducting devices*. J. Appl. Phys. **111**.5 (2012), p. 054510. DOI: `10.1063/1.3692073` (cit. on pp. 37, 38).

[Kir+13] G. Kirchmair, B. Vlastakis, Z. Leghtas, S. E. Nigg, H. Paik, E. Ginossar, M. Mirrahimi, L. Frunzio, S. M. Girvin, and R. J. Schoelkopf: *Observation of quantum state collapse and revival due to the single-photon Kerr effect*. Nature **495**.7440 (2013), pp. 205–209. DOI: `10.1038/nature11902` (cit. on p. 74).

[Kle15] K. Kleindienst: *Development and Design of a Bosonic Bath*. Bachelor thesis. (2015) (cit. on p. 109).

[Koc+07] J. Koch, T. M. Yu, J. Gambetta, A. A. Houck, D. I. Schuster, J. Majer, A. Blais, M. H. Devoret, S. M. Girvin, and R. J. Schoelkopf: *Charge-insensitive qubit design derived from the Cooper pair box*. Phys. Rev. A **76** (2007), p. 042319. DOI: `10.1103/PhysRevA.76.042319` (cit. on pp. 19–22, 27, 56, 60, 62, 78, 82).

[Koc+08] W. Koch, F. Großmann, J. T. Stockburger, and J. Ankerhold: *Non-Markovian Dissipative Semiclassical Dynamics*. Phys. Rev. Lett. **100** (2008), p. 230402. DOI: `10.1103/PhysRevLett.100.230402` (cit. on p. 99).

[KMS16] M. Koppenhöfer, M. Marthaler, and G. Schön: *Superconducting quantum metamaterials as an active lasing medium: Effects of disorder*. Phys. Rev. A **93** (2016), p. 063808. DOI: `10.1103/PhysRevA.93.063808` (cit. on p. 115).

[Kra16] P. Krantz: *The Josephson parametric oscillator – From microscopic studies to single-shot qubit readout*. PhD thesis. Quantum Device Physics Laboratory, Department of Microtechnology and Nanoscience, Chalmers University of Technology, 2016 (cit. on p. 37).

[Lan+17] N. K. Langford, R. Sagastizabal, M. Kounalakis, C. Dickel, A. Bruno, F. Luthi, D. J. Thoen, A. Endo, and L. DiCarlo: *Experimentally simulating the dynamics of quantum light and matter at deep-strong coupling*. Nat. Commun. **8**.1 (2017), p. 1715. DOI: `10.1038/s41467-017-01061-x` (cit. on pp. 74, 89).

[Lan+11] B. P. Lanyon, C. Hempel, D. Nigg, M. Müller, R. Gerritsma, F. Zähringer, P. Schindler, J. T. Barreiro, M. Rambach, G. Kirchmair, M. Hennrich, P. Zoller, R. Blatt, and C. F. Roos: *Universal Digital Quantum Simulation with Trapped Ions*. Science **334**.6052 (2011), pp. 57–61. DOI: `10.1126/science.1208001` (cit. on p. 4).

[Lan+10] B. P. Lanyon, J. D. Whitfield, G. G. Gillett, M. E. Goggin, M. P. Almeida, I. Kassal, J. D. Biamonte, M. Mohseni, B. J. Powell, M. Barbieri, A. Aspuru-Guzik, and A. G. White: *Towards quantum chemistry on a quantum computer*. Nat. Chem. **2**.2 (2010), pp. 106–111. DOI: `10.1038/nchem.483` (cit. on p. 2).

[Le 12] K. Le Hur: *Kondo resonance of a microwave photon*. Phys. Rev. B **85** (2012), p. 140506. DOI: `10.1103/PhysRevB.85.140506` (cit. on p. 119).

[Leg+87] A. Leggett, S. Chakravarty, A. Dorsey, M. Fisher, A. Garg, and W. Zwerger: *Dynamics of the dissipative two-state system*. Rev. Mod. Phys. **59**.1 (1987), pp. 1–85. DOI: `10.1103/revmodphys.59.1` (cit. on pp. 5, 99–101, 119).

[Leg+15] Z. Leghtas, S. Touzard, I. M. Pop, A. Kou, B. Vlastakis, A. Petrenko, K. M. Sliwa, A. Narla, S. Shankar, M. J. Hatridge, M. Reagor, L. Frunzio, R. J. Schoelkopf, M. Mirrahimi, and M. H. Devoret: *Confining the state of light to a quantum manifold by engineered two-photon loss*. Science **347**.6224 (2015), pp. 853–857. DOI: `10.1126/science.aaa2085` (cit. on p. 74).

[Lep+18] J. Leppäkangas, J. Braumüller, M. Hauck, J.-M. Reiner, I. Schwenk, S. Zanker, L. Fritz, A. V. Ustinov, M. Weides, and M. Marthaler: *Quantum simulation of the spin-boson model with a microwave circuit*. Phys. Rev. A **97** (2018), p. 052321. DOI: `10.1103/PhysRevA.97.052321` (cit. on pp. 5, 38, 99, 101, 103, 108–110, 115, 119).

[Li+13] J. Li, M. Silveri, K. Kumar, J.-M. Pirkkalainen, A. Vepsäläinen, W. Chien, J. Tuorila, M. Sillanpää, P. Hakonen, E. Thuneberg, and et al.: *Motional averaging in a superconducting qubit*. Nat. Commun. **4** (2013), p. 1420. DOI: `10.1038/ncomms2383` (cit. on p. 74).

[Lin76] G. Lindblad: *On the generators of quantum dynamical semigroups*. Commun. Math. Phys. **48**.2 (1976), pp. 119–130. DOI: 10.1007/BF01608499 (cit. on p. 9).

[Llo96] S. Lloyd: *Universal Quantum Simulators*. Science **273** (1996), pp. 1073–1078. DOI: 10.1126/science.273.5278.1073 (cit. on pp. 1, 2).

[Luc+12] E. Lucero, R. Barends, Y. Chen, J. Kelly, M. Mariantoni, A. Megrant, P. O'Malley, D. Sank, A. Vainsencher, J. Wenner, T. White, Y. Yin, A. N. Cleland, and J. M. Martinis: *Computing prime factors with a Josephson phase qubit quantum processor*. Nat. Phys. **8**.10 (2012), pp. 719–723. DOI: 10.1038/nphys2385 (cit. on pp. 2, 4).

[LU08] A. Lukashenko and A. V. Ustinov: *Improved powder filters for qubit measurements*. Rev. Sci. Instrum. **79**.1 (2008), p. 014701. DOI: 10.1063/1.2827515 (cit. on p. 48).

[Mag+18] L. Magazzú, P. Forn-Díaz, R. Belyansky, J.-L. Orgiazzi, M. A. Yurtalan, M. R. Otto, A. Lupascu, C. M. Wilson, and M. Grifoni: *Probing the strongly driven spin-boson model in a superconducting quantum circuit*. Nat. Commun. **9**.1 (2018), p. 1403. DOI: 10.1038/s41467-018-03626-w (cit. on p. 5).

[Mar+05] J. M. Martinis, K. B. Cooper, R. McDermott, M. Steffen, M. Ansmann, K. D. Osborn, K. Cicak, S. Oh, D. P. Pappas, R. W. Simmonds, and C. Y. Clare: *Decoherence in Josephson Qubits from Dielectric Loss*. Phys. Rev. Lett. **95**.21 (2005), p. 210503. DOI: 10.1103/physrevlett.95.210503 (cit. on p. 65).

[Mar+03] J. M. Martinis, S. Nam, J. Aumentado, K. M. Lang, and C. Urbina: *Decoherence of a superconducting qubit due to bias noise*. Phys. Rev. B **67**.9 (2003), p. 094510. DOI: 10.1103/physrevb.67.094510 (cit. on p. 63).

[Maz04] B. A. Mazin: *Microwave Kinetic inductance Detectors*. PhD thesis. California Institute of Technology, Pasadena, California, 2004 (cit. on p. 35).

[McC68] D. E. McCumber: *Effect of ac Impedance on dc Voltage-Current Characteristics of Superconductor Weak-Link Junctions*. J. Appl. Phys. **39**.7 (1968), pp. 3113–3118. DOI: 10.1063/1.1656743 (cit. on p. 19).

[Meg+12] A. Megrant, C. Neill, R. Barends, B. Chiaro, Y. Chen, L. Feigl, J. Kelly, E. Lucero, M. Mariantoni, P. J. J. O'Malley, D. Sank, J. Vainsencher, J. Wenner, T. C. White, Y. Yin, J. Zhao, C. J. Palmstrom, J. M. Martinis, and A. N. Cleland: *Planar superconducting resonators with internal quality factors above one million*. Appl. Phys. Lett. **100**.11 (2012), p. 113510. DOI: 10.1063/1.3693409 (cit. on p. 69).

[MO33] W. Meissner and R. Ochsenfeld: *Ein neuer Effekt bei Eintritt der Supraleitfähigkeit*. Naturwiss. **21**.44 (1933), pp. 787–788. DOI: 10.1007/BF01504252 (cit. on p. 13).

[MFK10] A. E. Miroshnichenko, S. Flach, and Y. S. Kivshar: *Fano resonances in nanoscale structures*. Rev. Mod. Phys. **82** (2010), pp. 2257–2298. DOI: 10.1103/RevModPhys.82.2257 (cit. on p. 110).

[Mos+16] S. Mostame, J. Huh, C. Kreisbeck, A. J. Kerman, T. Fujita, A. Eisfeld, and A. Aspuru-Guzik: *Emulation of complex open quantum systems using superconducting qubits*. Quantum Inf. Process. **16**.2 (2016), p. 44. DOI: 10.1007/s11128-016-1489-3 (cit. on p. 5).

[Mos+12] S. Mostame, P. Rebentrost, A. Eisfeld, A. J. Kerman, D. I. Tsomokos, and A. Aspuru-Guzik: *Quantum simulator of an open quantum system using superconducting qubits: exciton transport in photosynthetic complexes*. New J. Phys. **14**.10 (2012), p. 105013. URL: http://stacks.iop.org/1367-2630/14/i=10/a=105013 (cit. on pp. 2, 5).

[MCL17] C. Müller, J. H. Cole, and J. Lisenfeld: *Towards understanding two-level-systems in amorphous solids - Insights from quantum devices*. Preprint at https://arxiv.org/abs/1705.01108. (2017) (cit. on p. 65).

[Mül+15] C. Müller, J. Lisenfeld, A. Shnirman, and S. Poletto: *Interacting two-level defects as sources of fluctuating high-frequency noise in superconducting circuits*. Phys. Rev. B **92**.3 (2015), p. 035442. DOI: 10.1103/physrevb.92.035442 (cit. on p. 65).

[NPT99] Y. Nakamura, Y. A. Pashkin, and J. S. Tsai: *Coherent control of macroscopic quantum states in a single-Cooper-pair box*. Nature **398** (1999), pp. 786–788. DOI: 10.1038/19718 (cit. on p. 20).

[NC10] P. Nataf and C. Ciuti: *No-go theorem for superradiant quantum phase transitions in cavity QED and counter-example in circuit QED*. Nat. Commun. **1**.6 (2010), pp. 1–6. DOI: 10.1038/ncomms1069 (cit. on p. 96).

[Nee+09] M. Neeley, M. Ansmann, R. C. Bialczak, M. Hofheinz, E. Lucero, A. D. O'Connell, D. Sank, H. Wang, J. Wenner, A. N. Cleland, and et al.: *Emulation of a Quantum Spin with a Superconducting Phase Qudit*. Science **325**.5941 (2009), pp. 722–725. DOI: 10.1126/science.1173440 (cit. on p. 27).

[Nie+10] T. Niemczyk, F. Deppe, H. Huebl, E. P. Menzel, F. Hocke, M. J. Schwarz, J. J. Garcia-Ripoll, D. Zueco, T. Hümmer, E. Solano, A. Marx, and R. Gross: *Circuit quantum electrodynamics in the ultrastrong-coupling regime*. Nat. Phys. **6**.10 (2010), pp. 772–776. DOI: 10.1038/nphys1730 (cit. on pp. 5, 74).

[Nig+12] S. E. Nigg, H. Paik, B. Vlastakis, G. Kirchmair, S. Shankar, L. Frunzio, M. H. Devoret, R. J. Schoelkopf, and S. M. Girvin: *Black-Box Superconducting Circuit Quantization*. Phys. Rev. Lett. **108** (2012), p. 240502. DOI: 10.1103/PhysRevLett.108.240502 (cit. on p. 64).

[OIL13] P. Orth, A. Imambekov, and K. Le Hur: *Nonperturbative stochastic method for driven spin-boson model*. Phys. Rev. B **87**.1 (2013), p. 014305. DOI: 10.1103/physrevb.87.014305 (cit. on pp. 99, 119).

[PO10] H. Paik and K. D. Osborn: *Reducing quantum-regime dielectric loss of silicon nitride for superconducting quantum circuits*. Appl. Phys. Lett. **96**.7 (2010), p. 072505. DOI: 10.1063/1.3309703 (cit. on p. 66).

[Pai+11] H. Paik, D. I. Schuster, L. S. Bishop, G. Kirchmair, G. Catelani, A. P. Sears, B. R. Johnson, M. J. Reagor, L. Frunzio, L. I. Glazman, S. M. Girvin, M. H. Devoret, and R. J. Schoelkopf: *Observation of High Coherence in Josephson Junction Qubits Measured in a Three-Dimensional Circuit QED Architecture*. Phys. Rev. Lett. **107**.24 (2011). DOI: 10.1103/PhysRevLett.107.240501 (cit. on p. 54).

[Pal10] A. Palacios-Laloy: *Superconducting qubit in a resonator: Test of the Leggett-Garg inequality and single-shot readout*. PhD thesis. Quantum electronics group, CEA Saclay, 2010 (cit. on pp. 35, 37, 41).

[Par14] G. S. Paraoanu: *Recent Progress in Quantum Simulation Using Superconducting Circuits*. J. Low Temp. Phys. **175**.5-6 (2014), pp. 633–654. DOI: 10.1007/s10909-014-1175-8 (cit. on pp. 4, 53).

[Paz+16] G. A. Paz-Silva, S.-W. Lee, T. J. Green, and L. Viola: *Dynamical decoupling sequences for multi-qubit dephasing suppression and long-time quantum memory*. New J. Phys. **18**.7 (2016), p. 073020. DOI: 10.1088/1367-2630/18/7/073020 (cit. on p. 102).

[PA98] P. J. Petersan and S. M. Anlage: *Measurement of resonant frequency and quality factor of microwave resonators: Comparison of methods*. J. Appl. Phys. **84**.6 (1998), pp. 3392–3402. DOI: 10.1063/1.368498 (cit. on p. 37).

[Pis15] T. Piskor: *Design, Simulation and Development of a Bosonic Bath*. Bachelor thesis. (2015) (cit. on p. 110).

[PMO09] A. Politi, J. C. F. Matthews, and J. L. O'Brien: *Shoŕs quantum factoring algorithm on a photonic chip*. Science **325**.5945 (2009), p. 1221. DOI: 10.1126/science.1173731 (cit. on p. 2).

[Pop+14] I. M. Pop, K. Geerlings, G. Catelani, R. J. Schoelkopf, L. I. Glazman, and M. H. Devoret: *Coherent suppression of electromagnetic dissipation due to superconducting quasiparticles*. Nature **508** (2014), pp. 369–372. DOI: 10.1038/nature13017 (cit. on p. 54).

[Pot+18] A. Potočnik, A. Bargerbos, F. A. Y. N. Schröder, S. A. Khan, M. C. Collodo, S. Gasparinetti, Y. Salathé, C. Creatore, C. Eichler, H. E. Türeci, A. W. Chin, and A. Wallraff: *Studying light-harvesting models with superconducting circuits*. Nature Communications **9**.1 (2018), p. 904. DOI: `10.1038/s41467-018-03312-x` (cit. on p. 5).

[Poz12] D. M. Pozar: *Microwave Engineering*. 4th ed. Wiley, 2012. URL: `http://www.worldcat.org/isbn/0470631554` (cit. on pp. 29–35, 37, 38, 104).

[Pro+15] S. Probst, F. B. Song, P. A. Bushev, A. V. Ustinov, and M. Weides: *Efficient and robust analysis of complex scattering data under noise in microwave resonators*. Rev. Sci. Instrum. **86**.2 (2015), p. 024706. DOI: `10.1063/1.4907935` (cit. on pp. 37–39).

[qkit15] *Qkit – a quantum measurement and evaluation suite in python*. Karlsruhe Institute of Technology. URL: `https://github.com/qkitgroup/qkit` (cit. on p. 51).

[Hee] *QT Lab environment*. URL: `https://github.com/heeres/qtlab` (cit. on p. 51).

[Rab36] I. I. Rabi: *On the Process of Space Quantization*. Phys. Rev. **49**.4 (1936), pp. 324–328. DOI: `10.1103/physrev.49.324` (cit. on pp. 5, 24, 73).

[Rab37] I. I. Rabi: *Space Quantization in a Gyrating Magnetic Field*. Phys. Rev. **51**.8 (1937), pp. 652–654. DOI: `10.1103/physrev.51.652` (cit. on pp. 5, 24, 73).

[Ram49] N. F. Ramsey: *A new molecular beam resonance method*. Phys. Rev. **76**.7 (1949), p. 996. DOI: `10.1103/PhysRev.76.996` (cit. on p. 10).

[Ree+12] M. D. Reed, L. DiCarlo, S. E. Nigg, L. Sun, L. Frunzio, S. M. Girvin, and R. J. Schoelkopf: *Realization of three-qubit quantum error correction with superconducting circuits*. Nature **482**.7385 (2012), pp. 382–385. DOI: `10.1038/nature10786` (cit. on p. 4).

[Rei+17] M. Reiher, N. Wiebe, K. M. Svore, D. Wecker, and M. Troyer: *Elucidating reaction mechanisms on quantum computers*. Proc. Natl. Acad. Sci. USA **114**.29 (2017), pp. 7555–7560. DOI: `10.1073/pnas.1619152114` (cit. on p. 2).

[Rei+16] J.-M. Reiner, M. Marthaler, J. Braumüller, M. Weides, and G. Schön: *Emulating the one-dimensional Fermi-Hubbard model by a double chain of qubits*. Phys. Rev. A **94**.3 (2016), p. 032338. DOI: `10.1103/physreva.94.032338` (cit. on p. 70).

[Rig+12] C. Rigetti, J. M. Gambetta, S. Poletto, B. L. T. Plourde, J. M. Chow, A. D. Córcoles, J. A. Smolin, S. T. Merkel, J. R. Rozen, G. A. K. Keefe, M. B. Rothwell, M. B. Ketchen, and S. M.: *Superconducting qubit in a waveguide cavity with a coherence time approaching 0.1 ms.* Phys. Rev. B **86**.10 (2012), p. 100506. DOI: 10.1103/physrevb.86.100506 (cit. on p. 54).

[Ris+13] D. Ristè, C. C. Bultink, M. J. Tiggelman, R. N. Schouten, K. W. Lehnert, and L. DiCarlo: *Millisecond charge-parity fluctuations and induced decoherence in a superconducting transmon qubit.* Nat. Commun. **4** (2013), p. 1913. DOI: 10.1038/ncomms2936 (cit. on p. 62).

[Roc17] Private discussion with N. Roch, October 2017. (Cit. on p. 115).

[Sak94] J. J. Sakurai: *Modern Quantum Mechanics, Revised Edition.* Ed. by S. F. Tuan. Addison-Wesley Publishing Company, 1994 (cit. on pp. 9, 12, 24).

[Sal+15] Y. Salathé, M. Mondal, M. Oppliger, J. Heinsoo, P. Kurpiers, A. Potocnik, A. Mezzacapo, U. Las Heras, L. Lamata, E. Solano, S. Filipp, and A. Wallraff: *Digital Quantum Simulation of Spin Models with Circuit Quantum Electrodynamics.* Phys. Rev. X **5**.2 (2015), p. 021027. DOI: 10.1103/physrevx.5.021027 (cit. on p. 4).

[San+12] M. Sandberg, M. R. Vissers, J. S. Kline, M. Weides, J. Gao, D. S. Wisbey, and D. P. Pappas: *Etch induced microwave losses in titanium nitride superconducting resonators.* Appl. Phys. Lett. **100**.26 (2012), p. 262605. DOI: 10.1063/1.4729623 (cit. on p. 66).

[San+13] M. Sandberg, M. R. Vissers, T. A. Ohki, J. Gao, J. Aumentado, M. Weides, and D. P. Pappas: *Radiation-suppressed superconducting quantum bit in a planar geometry.* Appl. Phys. Lett. **102**.7 (2013), p. 072601. DOI: 10.1063/1.4792698 (cit. on pp. 54, 67).

[Sch97] V. V. Schmidt: *The Physics of Superconductors.* Ed. by P. Müller and A. Ustinov. Springer, 1997 (cit. on pp. 16, 17).

[Sch+12] U. Schneider, L. Hackermüller, J. P. Ronzheimer, S. Will, S. Braun, T. Best, I. Bloch, E. Demler, S. Mandt, D. Rasch, and A. Rosch: *Fermionic transport and out-of-equilibrium dynamics in a homogeneous Hubbard model with ultracold atoms.* Nat. Phys. **8**.3 (2012), pp. 213–218. DOI: 10.1038/nphys2205 (cit. on p. 4).

[SG08] R. J. Schoelkopf and S. M. Girvin: *Wiring up quantum systems.* Nature **451** (2008), pp. 664–669. DOI: 10.1038/451664a (cit. on pp. 4, 19).

[SSP14] M. Schuld, I. Sinayskiy, and F. Petruccione: *The quest for a Quantum Neural Network.* Quantum Inf. Process. **13**.11 (2014), pp. 2567–2586. DOI: 10.1007/s11128-014-0809-8 (cit. on p. 70).

[Sch07] D. I. Schuster: *Circuit Quantum Electrodynamics*. PhD thesis. Yale University, 2007 (cit. on pp. 25, 26, 35).

[Sha+11] S. Shahid, J. A. R. Ball, C. G. Wells, and P. Wen: *Reflection type Q-factor measurement using standard least squares methods*. IET Microw. Antennas Propag. **5**.4 (2011), pp. 426–432. DOI: 10.1049/iet-map.2010.0395 (cit. on p. 37).

[SMS02] A. Shnirman, Y. Makhlin, and G. Schön: *Noise and Decoherence in Quantum Two-Level Systems*. Phys. Scr. **2002**.T102 (2002), p. 147. URL: http://stacks.iop.org/1402-4896/2002/i=T102/a=024 (cit. on pp. 101, 103).

[Sho94] P. W. Shor: *Algorithms for quantum computation: discrete logarithms and factoring. Proceedings 35th Annual Symposium on Foundations of Computer Science*. 1994, pp. 124–134. DOI: 10.1109/SFCS.1994.365700 (cit. on p. 2).

[Sim01] R. N. Simons: *Coplanar waveguide circuits, components, and systems*. Ed. by K. Chang. Wiley series in microwave and optical engineering. John Wiley & Sons, 2001 (cit. on p. 31).

[SAD14] F. Solgun, D. W. Abraham, and D. P. DiVincenzo: *Blackbox quantization of superconducting circuits using exact impedance synthesis*. Phys. Rev. B **90**.13 (2014), p. 134504. DOI: 10.1103/physrevb.90.134504 (cit. on p. 64).

[Ste+06] M. Steffen, M. Ansmann, R. C., N. Bialczak, E. L. Katz, R. McDermott, M. Neeley, E. M. Weig, A. N. Cleland, and J. M. Martinis: *Measurement of the Entanglement of Two Superconducting Qubits via State Tomography*. Science **313**.5792 (2006), pp. 1423–1425. DOI: 10.1126/science.1130886 (cit. on pp. 4, 53).

[Ste16] A. Stehli: *Experiments with coupled concentric transmons*. MA thesis. Karlsruhe Institute of Technology (KIT), Physikalisches Institut, 2016 (cit. on pp. 69, 70, 117).

[Ste68] W. C. Stewart: *Current voltage characteristics of Josephson junctions*. Appl. Phys. Lett. **12**.8 (1968), pp. 277–280. DOI: 10.1063/1.1651991 (cit. on p. 19).

[Svo14] K. Svore: *Quantum Computing: Transforming the digital age*. 2014. URL: http://archive2.cra.org/uploads/documents/events/snowbird/2014slides/Quantum_Computing-Krysta_Svore.pdf (cit. on p. 2).

[Tim12] C. Timm: *Theory of Superconductivity*. Lecture notes, TU Dresden. (2012). URL: https://www.physik.tu-dresden.de/~timm/personal/teaching/thsup_w11/Theory_of_Superconductivity.pdf (cit. on pp. 13, 15).

[Tin04] M. Tinkham: *Introduction to Superconductivity, 2nd edition*. Dover, 2004 (cit. on pp. 13, 15, 17, 19).

[Tur37] A. M. Turing: *On Computable Numbers, with an Application to the Entscheidungsproblem. Proc. Lond. math. Soc. Ser. 2*. Vol. s2-42. 1. 1937, pp. 230–265. DOI: 10.1112/plms/s2-42.1.230 (cit. on p. 1).

[VEA12] S. Valleau, A. Eisfeld, and A. Aspuru-Guzik: *On the alternatives for bath correlators and spectral densities from mixed quantum-classical simulations*. J. Chem. Phys. **137**.22 (2012), p. 224103. DOI: 10.1063/1.4769079 (cit. on pp. 101, 102).

[Van+01] L. M. K. Vandersypen, M. Steffen, G. Breyta, C. S. Yannoni, M. H. Sherwood, and I. L. Chuang: *Experimental realization of Shorś quantum factoring algorithm using nuclear magnetic resonance*. Nature **414**.6866 (2001), pp. 883–887. DOI: 10.1038/414883a (cit. on p. 2).

[Vla+13] B. Vlastakis, G. Kirchmair, Z. Leghtas, S. E. Nigg, L. Frunzio, S. M. Girvin, M. Mirrahimi, M. H. Devoret, and R. J. Schoelkopf: *Deterministically Encoding Quantum Information Using 100-Photon Schrodinger Cat States*. Science **342**.6158 (2013), pp. 607–610. DOI: 10.1126/science.1243289 (cit. on pp. 4, 74).

[Voo+16] U. Vool, S. Shankar, S. O. Mundhada, N. Ofek, A. Narla, K. Sliwa, E. Zalys-Geller, Y. Liu, L. Frunzio, R. J. Schoelkopf, S. M. Girvin, and M. H. Devoret: *Continuous Quantum Nondemolition Measurement of the Transverse Component of a Qubit*. Phys. Rev. Lett. **117**.13 (2016), p. 133601. DOI: 10.1103/physrevlett.117.133601 (cit. on p. 77).

[Wal+04] A. Wallraff, D. I. Schuster, A. Blais, L. Frunzio, R.-. S. Huang, J. Majer, S. Kumar, S. M. Girvin, and R. J. Schoelkopf: *Strong coupling of a single photon to a superconducting qubit using circuit quantum electrodynamics*. Nature **431**.7005 (2004), pp. 162–167. DOI: 10.1038/nature02851 (cit. on pp. 23, 25, 78).

[Wei+11a] M. Weides, R. C. Bialczak, M. Lenander, E. Lucero, M. Mariantoni, M. Neeley, A. D. O'Connell, D. Sank, H. Wang, J. Wenner, Y. Yamamoto, Y. Yin, A. N. Cleland, and J. M. Martinis: *Phase qubits fabricated with trilayer junctions*. Supercond. Sci. Technol. **24**.5 (2011), p. 055005. DOI: 10.1088/0953-2048/24/5/055005 (cit. on p. 65).

[Wei+11b] M. P. Weides, J. S. Kline, M. R. Vissers, M. O. Sandberg, D. S. Wisbey, B. R. Johnson, T. A. Ohki, and D. P. Pappas: *Coherence in a transmon qubit with epitaxial tunnel junctions*. Appl. Phys. Lett. **99**.26 (2011), p. 262502. DOI: 10.1063/1.3672000 (cit. on p. 65).

[WS06] G. Wendin and V. Shumeiko: *Handbook of Theoretical and Computational Nanotechnology*. Ed. by M. Rieth and W. Schommers. Vol. 3. American Scientific Publishers, Los Angeles, 2006. Chap. Quantum Circuits, Qubits and Computing, pp. 223–309 (cit. on p. 8).

[Yos+16] F. Yoshihara, T. Fuse, S. Ashhab, K. Kakuyanagi, S. Saito, and K. Semba: *Superconducting qubit–oscillator circuit beyond the ultrastrong-coupling regime*. Nat. Phys. **13**.1 (2016), pp. 44–47. DOI: `10.1038/nphys3906` (cit. on pp. 5, 74).

[YN11] J. Q. You and F. Nori: *Atomic physics and quantum optics using superconducting circuits*. Nature **474** (2011), p. 589. DOI: `10.1038/nature10122` (cit. on p. 4).

[Zha+12] J. Zhang, M.-H. Yung, R. Laflamme, A. Aspuru-Guzik, and J. Baugh: *Digital quantum simulation of the statistical mechanics of a frustrated magnet*. Nat. Commun. **3** (2012), p. 880. DOI: `10.1038/ncomms1860` (cit. on p. 4).

List of publications

[Bra+15] J. Braumüller, J. Cramer, S. Schlör, H. Rotzinger, L. Radtke, A. Lukashenko, P. Yang, S. T. Skacel, S. Probst, M. Marthaler, L. Guo, A. V. Ustinov, and M. Weides: *Multiphoton dressing of an anharmonic superconducting many-level quantum circuit.* Phys. Rev. B **91**.5 (2015), p. 054523. DOI: 10.1103/physrevb.91.054523.

[Bra+17] J. Braumüller, M. Marthaler, A. Schneider, A. Stehli, H. Rotzinger, M. Weides, and A. V. Ustinov: *Analog quantum simulation of the Rabi model in the ultra-strong coupling regime.* Nat. Commun. **8**.1 (2017), p. 779. DOI: 10.1038/s41467-017-00894-w.

[Bra+16] J. Braumüller, M. Sandberg, M. R. Vissers, A. Schneider, S. Schlör, L. Grünhaupt, H. Rotzinger, M. Marthaler, A. Lukashenko, A. Dieter, A. V. Ustinov, M. Weides, and D. P. Pappas: *Concentric transmon qubit featuring fast tunability and an anisotropic magnetic dipole moment.* Appl. Phys. Lett. **108**.3 (2016), p. 032601. DOI: 10.1063/1.4940230.

[Lep+18] J. Leppäkangas, J. Braumüller, M. Hauck, J.-M. Reiner, I. Schwenk, S. Zanker, L. Fritz, A. V. Ustinov, M. Weides, and M. Marthaler: *Quantum simulation of the spin-boson model with a microwave circuit.* Phys. Rev. A **97** (2018), p. 052321. DOI: 10.1103/PhysRevA.97.052321.

[Rei+16] J.-M. Reiner, M. Marthaler, J. Braumüller, M. Weides, and G. Schön: *Emulating the one-dimensional Fermi-Hubbard model by a double chain of qubits.* Phys. Rev. A **94**.3 (2016), p. 032338. DOI: 10.1103/physreva.94.032338.

Appendix

Fabrication parameters

Table 1: Substrate cleaning

solvent	ultrasonic bath	IPA	H_2O	hot plate	piranha rinse
NEP	10 min	yes	yes	145 °C, 1 min	yes

Electron beam lithography

Table 2: Application of the resist stack

resist	pre	spin	hotplate
LOR 500nm	300 rpm, 10 s	3000 rpm, 60 s	180 °C, 5 min
PMMA 950K AR-P672.03	300 rpm, 10 s	6000 rpm, 60 s	145 °C, 5 min

Table 3: Lithography with a 10 keV *Jeol* electron beam writer

	write current	dose
large structures, qubit electrodes	6 nA	350 μCcm^{-2}
junction leads, bridge	100 pA	300 μCcm^{-2}

Table 4: Development

developer	time	rinse	dry
MIBK/IPA 1:3, 25 °C	65 s	IPA 30 s, H_2O 60 s	spin coater, 7500 rpm, 120 s
MIF 726/H_2O 3:2	150 s	H_2O 60 s	spin coater, 7500 rpm, 120 s

Table 5: Plasma clean with the Kaufman source, *Plassys*

plasma voltage	current	gas	time
200 V	10 mA	4 sccm Ar, 0.5 sccm O_2	10 s

Table 6: Al Shadow angle evaporation, *Plassys*

	thickness	rate	angle
Al layer 1	30 nm	0.2 nm/s	8°
Al layer 2	50 nm	0.2 nm/s	–8°

Table 7: Dynamic oxidation between Al layers 1,2

O_2 flow rate	pressure	time
12.75 sccm	0.021 mbar	45 min

Table 8: Lift-off

solvent	hotplate	time
NEP	75 °C	$\sim$ 1 h

Optical lithography

Table 9: Primer application as sticking layer for photo-resist

agent	hotplate	pump in desiccator
HMDS	180 °C	2 min + 5 min

Table 10: Resist application

resist	pre	spin	hotplate
AZ 5214E	500 rpm, 5 s	6000 rpm, 60 s	110 °C, 50 s

Table 11: Lithography at a *Carl Süß* mask aligner, 500 W XeHg lamp

dose	exposure time	mask type
$13\,mW(cm)^{-2}$	5 s	soda lime, hard contact mode

Table 12: Development

developer	time	rinse
MIF 726	35 s	H_2O
AZ developer, H_2O 1:1	33 s	H_2O

Table 13: Plasma clean with the Kaufman source, *Plassys*

plasma voltage	current	gas	time
200 V	10 mA	4 sccm Ar, 0.5 sccm O_2	10 s

Table 14: AuPd evaporation, *Plassys*

thickness	rate	contact layer
75 nm	$\sim 0.2\,nm/s$	5 nm titanium

Table 15: Plasma clean in *Plasma 1*

plasma power	voltage	gas	time	comment
20 W	80 V	10 sccm Ar	2 min	prior to Cu deposition

Table 16: Sputter deposition, *Plasma 1*

target	sputter power	Ar flow rate	main chamber pressure	time
Cu	130 W	30 sccm	1.9×10^{-3} mbar	105 s

Table 17: Lift-off

solvent	ultrasonic bath	time
NEP	power 5/9	$\sim$ 5 min to 15 min

Acknowledgements

I want to express my gratitude to Prof. Alexey Ustinov for giving me the opportunity to perform this Ph.D. work under his supervision. Throughout my time in his group he generously provided me with the possibility to attend several conferences, schools and workshops. These have significantly helped me to become familiar with the scientific community in the field of superconducting quantum circuits and have led to the development of fruitful collaborations and good friendships. I appreciate the scientific freedom I had, which allowed me to pursue and realize own ideas. I want to thank Alexey for making possible the annual winter group seminars at cold temperatures and for the two amazing (sailing) seminars on high frequency waves.

My gratitude goes to Prof. Gerd Schön for accepting to be my co-examiner. I am thankful for constructive criticism and theoretical input in particular during the last weeks before my defence. Additionally, I appreciate his valuable and helpful comments on this thesis.

I am thankful to Dr. Martin Weides for advising me during this thesis. He was originally fostering the project of quantum simulation and established the contact to Prof. Dave Pappas' group at NIST which largely influenced this thesis. I am grateful for many sessions of brainstorming and the financial means he provided. Thank you also for initiating and enabling the research exchange with Prof. Arkadi Fedorov's lab in Brisbane.

I want to thank Dr. Michael Marthaler for our very close collaboration throughout the course of my thesis. He provided crucial advice and suggestions not only from the theory side that decisively influenced the direction of this thesis. We have an ongoing bet on the coherence of concentric transmon qubits, which made me swim in the Fermi lake by the physics building already some time ago. It is a pending achievement to further enhance the coherence (by few microseconds) and have Michael try the refreshing lake.

I am thankful to Dr. Juha Leppäkangas for being an extremely valuable theory support during the last part of my thesis. He helped me to become acquainted with the basic theoretical concepts of the spin boson model and provided important numerical simulations.

My thanks go to Dr. Hannes Rotzinger for his everyday help with any possible problem on physics and beyond. My acquired knowledge about the linux kernel and Python coding skills is largely due to him.

I want to thank Dr. Ioan Pop for very valuable support in optimizing my talks, in particular for the APS spring meeting 2017. In addition, I am thankful for various stimulating physics discussions.

I want to thank Dr. Jürgen Lisenfeld for a critical proof reading of parts of this thesis and for valuable expert opinions and advice on qubit measurements and two-level systems in general.

A big thanks goes to Dr. Sasha Lukashenko for his invaluable support operating the 'UFO' cryostat. He was taking night and weekend shifts when a blocked capillary was preventing us from cooling down and could solve the problem by an incredible solder job.

I want to thank Lucas Radtke for his constant support in sample fabrication and his help with any question arising in the clean room. He continuously pushed the levels in optical lithography, making it very reliably working. My thanks also go to Dr. Silvia Diewald for operating the electron beam writer and for an efficient cooperation in optimizing the dose pattern for proximity correction.

Many thanks go to Prof. Arkadi Fedorov for supervising my research exchange attendance in his group in Brisbane, Australia. I could learn a lot about two-level fluctuators and superconducting circuits from Dr. Markus Jerger, Dr. Clemens Müller and Dr. Pascal Macha.

I want to acknowledge financial support by the Landesgraduiertenförderung (LGF) of the federal state Baden-Württemberg and by the Helmholtz International Research School for Teratronics (HIRST).

Ich möchte mich bei der großartigen Unterstützung durch die Werkstätten des Physikalischen Instituts bedanken. Die Zusammenarbeit mit Roland Jehle und Michael Meyer war stets sehr effektiv und lehrreich. Des Weiteren gilt mein Dank Hans-Willi Pensl für immer ausreichenden Helium-Nachschub und ein unkompliziertes spontanes Befüllen fast leerer Kannen. Ich bedanke mich außerdem bei Dr. Christoph Sürgers für Rat und Tat wann immer nötig.

Mein Dank geht auch an Steffi Baatz, Ilona Brosch und Birgit Schelske für Ihre Hilfe bei jeder organisatorischen Frage und Ihrer Geduld bei komplizierten Urlaubsanträgen.

Ein besonderer Dank gilt den Studenten die ich während meiner Promotion betreut habe und die Ihren Teil zum Gelingen dieser Arbeit beigetragen haben. Dies waren

Bachelorarbeiten von Kai Kleindienst und Tomislav Piskor, sowie studentische Praktika von Clark Baker und Laith Meti. Insbesondere möchte ich mich für die großartige Zusammenarbeit mit Alexander Stehli im Zuge seiner Masterarbeit bedanken. Überdies schätze ich unsere zahlreichen physikalischen Diskussionen und andere Momente in denen 'Vollgas' gegeben wurde.

Ich möchte mich bei Dr. Markus Jerger bedanken für Bau und Programmierung des 'Time domain setup'. Dies war eine wertvolle Grundlage für Weiterentwicklungen und Messungen in dieser Arbeit. Mein Dank gilt auch Dr. Sebastian Probst für seine Hilfe und Expertise insbesondere zu Mikrowellenmessungen und dem 'Cricle fit'. Ich möchte mich außerdem bedanken bei Sebastian Skacel für wertvolle Ratschläge vor allem bei Fabrikationsfragen und der Erstellung von Vorträgen. Mein Dank geht auch an Patrick Winkel für professionelle Photografien und deren graphische Nachbearbeitung.

Mein Dank gilt dem 'UFO Team' für erfolgreiche Zusammenarbeit beim Betrieb des Kryostats. Neben Dr. Sasha Lukashenko sind dies Steffen Schlör und Andre Schneider, sowie Alexander Stehli, Ping Yang, Tim Wolz und Jan Brehm. Außerdem bedanke ich mich bei meinen Büro-Kollegen Sebastian Skacel, Marco Pfirrmann und Alexander Bilmes für eine stets heitere Atmosphäre und fachliche Hilfsbereitschaft. Ich danke der gesamten Arbeitsgruppe einschließlich Untergruppen für die positive Stimmung sowie tolle Erlebnisse auf Konferenzen.

Mein Dank geht an Andre Schneider, Alexander Stehli, Dr. Jürgen Lisenfeld, Dr. Martin Weides, Dr. Hannes Rotzinger und Dr. Michael Marthaler für das Korrekturlesen meiner Arbeit.

Schließlich möchte ich mich bei meiner Mutter bedanken für Ihre Unterstützung während dieser Arbeit und dafür dass Sie immer an mich geglaubt hat. Mein besonderer Dank geht an Christina, die immer zu meiner Seite stand. Sie hatte ein offenes Ohr für allerlei Probleme und es gelang Ihr stets mich aufzuheitern wenn nötig.

Karlsruhe, November 2017 Jochen Braumüller

Experimental Condensed Matter Physics
(ISSN 2191-9925)

Herausgeber
Physikalisches Institut

Prof. Dr. David Hunger
Prof. Dr. Alexey Ustinov
Prof. Dr. Georg Weiß
Prof. Dr. Wolfgang Wernsdorfer
Prof. Dr. Wulf Wulfhekel

Band 1 Alexey Feofanov
Experiments on flux qubits with pi-shifters. 2011
ISBN 978-3-86644-644-1

Band 2 Stefan Schmaus
Spintronics with individual metal-organic molecules. 2011
ISBN 978-3-86644-649-6

Band 3 Marc Müller
Elektrischer Leitwert von magnetostriktiven Dy-Nanokontakten. 2011
ISBN 978-3-86644-726-4

Band 4 Torben Peichl
Einfluss mechanischer Deformation auf atomare Tunnelsysteme – untersucht mit Josephson Phasen-Qubits. 2012
ISBN 978-3-86644-837-7

Band 5 Dominik Stöffler
Herstellung dünner metallischer Brücken durch Elektromigration und Charakterisierung mit Rastersondentechniken. 2012
ISBN 978-3-86644-843-8

Band 6 Tihomir Tomanic
Untersuchung des elektronischen Oberflächenzustands von Ag-Inseln auf supraleitendem Niob (110). 2012
ISBN 978-3-86644-898-8

Band 7 Lukas Gerhard
Magnetoelectric coupling at metal surfaces. 2013
ISBN 978-3-7315-0063-6

Band 8 Kirill Fedorov
Fluxon readout for superconducting flux qubits. (nicht erschienen)
ISBN 978-3-7315-0067-4

Die Bände sind unter www.ksp.kit.edu als PDF frei verfügbar oder als Druckausgabe bestellbar.

Band 9 Jochen Zimmer
Cooper pair transport in arrays of Josephson junctions. 2014
ISBN 978-3-7315-0130-5

Band 10 Oliver Berg
Elektrischer Transport durch Nanokontakte von Selten-Erd-Metallen. 2014
ISBN 978-3-7315-0209-8

Band 11 Grigorij Jur'evic Grabovskij
Investigation of coherent microscopic defects inside the tunneling barrier of a Josephson junction. 2014
ISBN 978-3-7315-0210-4

Band 12 Cornelius Thiele
STM Characterization of Phenylene-Ethynylene Oligomers on Au(111) and their Integration into Carbon Nanotube Nanogaps. 2014
ISBN 978-3-7315-0235-7

Band 13 Michael Peter Schackert
Scanning Tunneling Spectroscopy on Electron-Boson Interactions in Superconductors. 2014
ISBN 978-3-7315-0238-8

Band 14 Susanne Butz
One-Dimensional Tunable Josephson Metamaterials. 2014
ISBN 978-3-7315-0271-5

Band 15 Philipp Jung
Nonlinear Effects in Superconducting Quantum Interference Meta-Atoms. 2014
ISBN 978-3-7315-0294-4

Band 16 Sebastian Probst
Hybrid quantum system based on rare earth doped crystals. 2016
ISBN 978-3-7315-0345-3

Band 17 Wolfram Kittler
Magnetische Anisotropie und Quantenphasenübergang in $CeTi_{1-x}V_xGe_3$. 2015
ISBN 978-3-7315-0363-7

Band 18 Moritz Peter
Towards magnetic resonance in scanning tunneling microscopy using heterodyne detection. 2015
ISBN 978-3-7315-0410-8

Die Bände sind unter www.ksp.kit.edu als PDF frei verfügbar oder als Druckausgabe bestellbar.

Band 19 Junji Tobias Märkl
Investigation of Magnetic Adatoms with Scanning Tunneling Techniques. 2015
ISBN 978-3-7315-0435-1

Band 20 Jochen Braumüller
Quantum simulation experiments with superconducting circuits. 2018
ISBN 978-3-7315-0780-2